A Story of Dreams, Doubts, Faith, and Triumph

I Followed
A
Different Star

The Memoirs of

Geoffrey Yoder

From Farm Boy to NASA Executive

Editor: Rachel Arterberry at Making a Way Writing Services www.makinga waywriting.com

eBook ISBN: 979-8-9987404-2-8

Paperback : 979-8-9987404-3-5

Hardcover: 979-8-9987404-4-2

Cover Photo: Cosmic Wreath as modified by Geoffrey Yoder

Photo Credit: NASA/JPL X-ray: NASA/CXC; Infrared: ESA/Webb, NASA & CSA, P. Zeilder, E. Sabbi, A. Nota, M. Zamani; Image Processing: NASA/CXC/SAO/L. Frattare and K. Arcand

Praise for I Followed a Different Star

"With more than 16 years in the industry and 16 years at NASA, Geoff's story is not only of individual success and hard work, but also of NASA's transition to a new era of space exploration, in which he played many key leadership roles. He has accomplished what most of us come here hoping to do—move our mission—and America's space program—forward." *Former NASA Administrator, Retired Major General Charles F. Bolden Jr.*

"In the profound stillness of the midnight hour, above the sun-drenched fields of his family farm, Geoff found his truest calling. Chronicled in the inspiring narrative "I Followed a Different Star," his journey began beneath a velvet cloak studded with diamond dust. For Geoff, born into a world dictated by the rhythm of plows, the celestial expanse was not merely a distant spectacle but a vibrant, breathing entity—a silent, infinite canvas that painted dreams upon his young, impressionable mind. Destined, by popular decree, to be nothing more than a farmer, the midnight sky whispered of possibilities far grander than any earthly furrow.

This remarkable ascent, an improbable journey from the fields of his youth to the gleaming, high-tech corridors of NASA, masterfully illustrates how a young man, told he wouldn't amount to anything beyond his inherited plot of land, dared to look up and dream beyond the immediate horizon. Fueled by an unquenchable curiosity, the unwavering support of family and coworkers, and a relentless, almost fierce dedication to unraveling the cosmos's secrets, Geoff's path was one of audacious ambition against formidable odds.

His gripping narrative culminates in astonishing achievement, a testament not only to raw intellect and unwavering determination but also to the audacious pursuit of a seemingly impossible dream. Reaching the pinnacle of his professional aspirations within the hallowed halls of scientific innovation, Geoff's story delves deeper than mere professional triumph. The book profoundly explores the 'how' and, crucially, the 'why' of his success, illuminating the intrinsic role of his unshakeable faith.

His spiritual convictions were not just a guiding light from the periphery but an active, permeating force woven into the very fabric of his challenging journey. As "I Followed a Different Star" beautifully illustrates, a life dedicated to meticulous scientific exploration can coexist with, and indeed be profoundly enriched by, an abiding faith. For Geoff, the pursuit of knowledge and the deepening of his spiritual life became two inseparable sides of the same incredible quest, each fueling a deeper sense of wonder and purpose.

This poignant and comprehensive portrait of a life lived with passion, purpose, and profound belief reminds us that the greatest journeys are often those that transcend earthly boundaries, touching the very fabric of existence and revealing the boundless possibilities when one dares to follow a different star." *Kaya Jones, artist, singer/songwriter, and Grammy Award winner*

Dedication

W ith heartfelt appreciation, I dedicate this work to my cherished family: my wife, son, daughter-in-law, and grandson.

Eternal thank-you to my mother who did not give up on me in my early years of questioning the meaning and purpose of life.

Beyond the personal realm, I extend my profound gratitude to the NASA leadership, whose unwavering trust in my ability to lead various organizations has been both an immense honor and a significant driving force.

Finally, a special appreciation for all the dedicated teachers and educators who tirelessly cultivate the minds of our youth, shaping future generations with their wisdom and passion. And to every reader embarking on their own journey, may this book serve as a beacon of hope, reminding you to pursue your dreams relentlessly, even when the path seems impossible to navigate.

With best regards,
Geoff Yoder

Contents

Reflections

I found myself sitting in a surprisingly quiet hotel restaurant in Germany, the evening light filtering gently through the window. A warm cup of tea was steaming in front of me, and as I took a slow sip, a wave of genuine amusement washed over me. Here I was, little old me, a farm boy at heart, now a senior executive at NASA. My schedule for the series of days? Crucial discussions with representatives from thirteen different space agencies, all gathered to talk about creating a framework for exploration. The sheer improbability of it all was not lost on me; in fact, it was a moment that made me chuckle inwardly. How in the world did I end up here?

My journey had not been paved with academic accolades or effortless success. Far from it. In fact, my early academic years were marked by struggle, confusion, and, at times, a deeply disheartening sense of discouragement. I remember, quite vividly, the abstract world of mathematics feeling incredibly irrelevant by the time I hit fourth grade, resulting in a disheartening score that left me feeling rather deflated. It wasn't just math, either. In seventh grade, I had a teacher who, despite my polite corrections, persistently mispronounced my name. It chipped away at my confidence, day by day, fostering a negative attitude toward education in general. Why try, I wondered, when it felt like I wasn't even seen or heard correctly?

High school brought more of the same. My algebra and physics teachers were often visibly frustrated with me, convinced I wasn't giving it my all, that I was simply being lazy. I'm sure, looking back, that my performance certainly made it look that way. I was disengaged, the subjects felt disconnected from anything I cared about, and I simply couldn't see the point. Then came my junior year when one of my teachers delivered a harsh assessment: "You grew up on a farm; you'll never be anything more than a farmer."

They were like seeds, planting themselves in the fertile ground of my nascent self-doubt and insecurity. They threatened to define my potential, to confine me to a reality I wasn't sure I wanted. Would this really be my destiny? Was I truly limited by my origins, by the dust and soil of the farm? It was a question that haunted me for a time, a heavy weight on my young shoulders.

However, I was incredibly fortunate to have a strong support system —and I truly can't emphasize enough how vital this was. My friends, my family, and, as you'll hear, a particularly influential coworker later in my life, all played crucial roles. They helped me realize that other people's opinions, no matter how forcefully delivered, don't have to dictate my destiny. They reminded me that I had the power to write my own story. It wasn't an instant revelation, but a gradual awakening, nurtured by their unwavering belief in me.

The road from that realization to where I am today was far from easy. In fact, it required hard work and an unwavering, stubborn perseverance. There were countless late nights studying, moments of frustration when I wanted to give up, and the constant battle against that lingering voice of doubt, often my own. I had to fundamentally change my approach to learning, to embrace disciplines that once felt like insurmountable obstacles. It meant reaching out for help, asking questions, and slowly, painstakingly, building a foundation of knowledge and confidence that had been missing.

But through these efforts, something remarkable happened. I not only met but surpassed my own expectations. From herding cows and fixing tractors in my youth, with dirt under my fingernails and the smell of hay in my hair, I eventually found myself overseeing all of NASA's Civilian Space Science. Think about that for a moment: *all* of it. This included the awe-inspiring projects of Astrophysics, like the incredible Hubble and James Webb Space Telescopes, peering back in time to the very beginnings of our universe. It encompassed Planetary Science, exploring distant worlds within our solar system, from Mars to the outer gas giants. Earth Science, understanding our own precious planet from a cosmic perspective, and Heliophysics, delving into the mysteries of our sun.

Through sharing my story, delving into the highs and lows you will see that it is more than just a collection of chapters; it's a map of a life lived, etched with the indelible marks of both soaring triumphs and gut-wrenching falls. As I poured my story onto those pages, delving deep into the exhilarating highs and the humbling lows, one truth became clearer than ever, a truth I pray echoes in the heart of anyone who reads or hears it: It is not only acceptable to face doubts and setbacks, but it is profoundly *essential.*

Many might expect a narrative interwoven with faith to present a path devoid of struggle, a life shielded by divine intervention. But my journey, as faithfully recounted, reveals something far more resonant. My deeply held faith isn't a magic cloak that made problems vanish; instead, it was the compass that helped me navigate the storm, the quiet strength that whispered reassurance when the winds raged strongest. It's a faith that doesn't promise an easy road, but guarantees a steadfast companion *on* the road, particularly when the path grows treacherous.

Those moments of doubt, those unexpected setbacks that felt like concrete walls slamming shut, were never presented in my story as failures. They were, and are, integral parts of the journey. I've come to see them not as roadblocks designed to halt progress, but as challenging terrain. Imagine a mountain trail – the flat, easy stretches are pleasant, but it's the steep inclines, the rocky paths, the winding descents that truly test your mettle. It's through navigating that difficult terrain that your muscles strengthen, your resolve hardens, and your perspective widens.

This is the crucial message I hope to instill: these challenging moments do not diminish you. They are not the final word on who you are or what you can achieve. They don't define you; your response to them does. Do you crumble, or do you seek the lesson, find the resilience, and push forward, perhaps a little wiser, certainly a little stronger? My story, with all its imperfections and triumphs, is a testament to the fact that embracing the full spectrum of the human experience – the glorious and the grimy – is where true growth lies. May you, too, find the courage to see your challenges not as walls, but as the very ground upon which your strength is forged.

When the path ahead seems unclear, overwhelming, or daunting, I want to inspire you to persevere. I want you to remember my story, a boy told he'd never amount to more than a farmer, discussing the future of space exploration with international leaders, and ultimately leading the NASA Science Mission Directorate. You possess an inherent, incredible power to shape your own path to success and fulfillment, regardless of how intimidating it may seem at first glance. Embrace the possibilities that lie before you, believe in yourself with all your heart, and, truly, reach for the stars. They might just be closer than you think.

Chapter One

The Early Years

My Farm Life Adventures (and Misadventures!)

Growing up on a small dairy farm in western Pennsylvania was quite the ride. I'm not talking about some idyllic, picture-perfect childhood where everything was sunshine and rainbows. Nope, this was real life, complete with rising before the crack of dawn, wrestling with grumpy cows, shoveling feed, and attempting to mend fences that always seemed to have a mind of their own. You name it, we probably did it. It wasn't just a list of chores; it was like a masterclass in hard work, responsibility, and ingenuity, all packed into my early childhood. I learned how to drive a tractor before I could even properly reach the pedals, and let me tell you, those countless hours I spent playing in the barn and fields? Looking back, that was way cooler than being glued to any screen. Zero regrets there.

Photo of the family farm in the 1970's

Every single morning and evening, like clockwork, it was our job to bring those cows in from the fields to be milked. We'd also feed them grain in the barn and set out silage or fodder for them to munch on. Most of the time, those cows were practically waiting at the gate, super eager to be milked. But then there were *those* days—oh, those days—when they'd decide to just hang out at the far end of the field, being totally uncooperative. Instead of trudging through all that tall, wet grass to round them up, I would hop on my trusty Honda 350 motorcycle. Seemed like a smart move at the time, right? Efficient, even.

Well, one time, I was cruising through that tall grass, feeling all smug and efficient, and completely missed seeing a huge hole in the ground. I hit it dead center. Next thing I knew, I was flying over the handlebars, landing with a rather unpleasant thump on the grass while my motorcycle tumbled right after me. The engine was still running, almost like it was laughing at my misfortune. I gingerly managed to get back on the bike, my ankle throbbing like crazy, and rode home. The worst part? Dreading the task of telling my dad what had happened. Of course, he just thought I was trying to get out of work. He reminded me that when he was young, he had horses to gather the cattle and certainly didn't need a fancy motorcycle.

His comment about horses immediately zipped my mind back to several years earlier when I had a rather unique experience with a horse that frightened me so much, I swore I would never get on one again. I was quite young when the horse I was riding was stopped by a closed fence. This particular horse, apparently not a fan of waiting, decided to take off on its own toward another open gate. And well, I slid off the saddle, only to have my foot stuck in the saddle stirrup. So, there I was, dangling under the horse as it galloped toward the other gate! It was only when we reached the gate that my foot finally came loose from the stirrup and I fell to the ground. That was the moment I definitively decided that driving a motorcycle, where I felt like I "controlled" the outcome, was way better than being on a horse with a mind of its own.

Farm life was a wild mix of hard work, unexpected adventures, and definitely a few bumps and bruises. But honestly, looking back, I wouldn't trade those experiences for anything. They weren't just stories; they were lessons that stuck.

That motorcycle I was riding to herd cattle. I actually got it when I was 16. I saw an ad for a Honda 350 in the local newspaper and decided to check it out. Let me tell you, it was a mess. The chassis was covered in oil, the drive chain was dry, the engine sputtered, and the exhaust pipes were coated with an oily residue. It clearly hadn't been loved in a while. I didn't pay much for it, given its poor condition, but I knew I could bring it back to life. I scrubbed off the grease, tightened the drive chain, and made the muffler shine like new. After the test ride, I parked it in the garage. That's when my little brother, who was probably around 5 or 6, spotted the shiny area where the tailpipe leaves the engine and, for some reason, decided it was a good idea to give it a big smooch. Yeah, lessons were learned that day, not just by me!

Working on a farm was no joke—there were so many routine chores to be done every single day! From sunup to sundown, there was always something that needed doing: milking the cows, cleaning out the stables, fixing fences, planting crops for the upcoming harvest, and a whole lot more. It was relentless.

But amidst all that hard work, we had our simple pleasures. I remember going to a local ice cream parlor with my dad after a long day in the fields. Those ice cream cones were a special treat and definitely didn't happen very often, which made them all the more delicious. During the summer months, our lives revolved around the farm; we were pretty isolated from the rest of society, except for weekends when we'd attend church. The days were long and hot, but there was a certain satisfaction, a deep feeling of accomplishment, in seeing the fruits of our labor. At the end of the day, we were physically tired, sure, but emotionally charged and surprisingly content.

I spent most of my time alone, just listening to the sounds of nature: the crickets chirping, leaves rustling, and the occasional moo from our livestock. It was a peaceful way of life, filled with hard work and those simple joys like watching the sunset over the fields. Sure, I missed having company sometimes, but there was something special about feeling so connected to the land and all its changes throughout the year. It built a strong sense of self-reliance, that's for sure.

Growing up in a low-income farm family as the middle child of eight siblings sure had its challenges. Not only did we have to deal with limited money, but I always felt like I was stuck in the middle—not the oldest or youngest, not the loudest or quietest. It was tough trying to stand out and make a name for myself when I was constantly competing for attention among so many siblings. And the physical demands of farm life left barely any time for fun, which added to the challenge. Plus, my older brother suffered from severe allergies, so he couldn't help harvesting hay, which dumped even more workload on the rest of us.

Growing up with siblings is a total mixed bag. On one hand, you've got a constant source of entertainment (or, let's be honest, occasional torture). On the other hand, well, sometimes the torture definitely outweighed the entertainment. I remember being around 5 or 6 years old, and my siblings would get their kicks out of teasing me.

They'd pick on me, then whisper, "Don't tell Mom and Dad!" Classic sibling maneuver.

But kids are resilient, and I quickly learned to adapt. My escape? A good old-fashioned imaginary world, deeply inspired by the farm we lived on. You see, our barn had cows, and those cows were often tied to things (presumably to keep them from wandering off and causing bovine mayhem). For some reason, this image really stuck with me.

So, what did I do? I would sneak into the kitchen, find a sturdy chair to hide under, and then tie myself to it with whatever I could find—string, ribbon, maybe even a rogue shoelace. My rationale? I was pretending to be a cow. And the space under the kitchen chair was my "safe space," a quiet space where I could escape the sibling shenanigans.

Of course, my idyllic bovine fantasy didn't last long. My siblings, naturally, discovered my hiding spot. And did they leave me alone to moo in peace? Absolutely not. The teasing just evolved. Now, instead of just being teased, I was being teased for pretending to be a cow tied to a chair.

Looking back, it is kind of hilarious. It is a quirky memory that paints a vivid picture of childhood imagination, sibling rivalry, and the desperate need for a little peace and quiet.

While I may not be tying myself to chairs anymore, the memory serves as a reminder that even amidst the chaos of family life, finding your own "safe space," even if It is under a kitchen chair pretending to be a cow, is essential.

Another memory that stands out is when my family took a trip to Somerset, PA, about 30 miles from our home. My parents had to attend to some paperwork at the courthouse, so my siblings and I stayed in the car for about half an hour while they were inside. I was probably seven or eight at the time, and, of course, my siblings could not resist picking on me. They decided to make up this super spooky story just to scare me.

They noticed smoke coming from the courthouse chimney and concocted this wild tale about them burning naughty children inside. They even told me I had siblings named Orfaorfa and Jerryberry (yeah, totally made up, obviously). They claimed the smoke was from my siblings being burned for misbehaving and that our parents were inside dealing with my behavior for a possible burning, too. Siblings can be so mean sometimes! As my parents approached the car, my siblings warned me not to spill the beans about our conversation or else they'd hurt me. Oh, the things siblings do to mess with you!

How My Childhood Shaped Me

Life on the farm wasn't always a walk in the park. We had our fair share of struggles, but looking back, those formative years were absolutely priceless. They hammered home some serious lessons about hard work, perseverance, and that unbreakable bond you only get with family.

Being the middle child, I often felt like I was vying for attention, sometimes getting a bit lost in the bustling chaos of my older and younger siblings. But in those moments when I felt overlooked, I found solace in the most unlikely companion: our family dog. Whether we were chasing after a beat-up tennis ball under the scorching sun or just sharing a quiet moment of companionship, my furry friend was always there. My dreams even ventured into canine territory, fantasizing about one day running my very own German Shepherd farm—a testament to the profound joy and comfort a simple animal can bring, especially when life throws you a curveball.

But not every dog encounter was a heartwarming tale. I vividly recall the sheer terror when our neighbor's German Shepherd ventured onto our property. Its aggressive behavior would send me scrambling for the nearest refuge, which was usually a sturdy, climbable tree. Forget running back to the house; that felt way too predictable, and honestly, a bit too slow! Inspired by stories like "Peter and the Wolf," where a brave boy outsmarts a wolf with cunning and a well-placed rope, I decided to arm myself. I stashed a rope in my trusty tree, ready to lasso should the canine menace reappear. While my attempts at lassoing the aggressive dog never quite matched Peter's heroic feat, the thrill of the challenge and the feeling of taking control were undeniably exhilarating. Who doesn't love a good underdog story, even if the "underdog" is a kid in a tree with a rope?

Beyond the canine adventures, my childhood wardrobe also told a story of resourcefulness and a whole lot of love. New clothes were a luxury rarely afforded by our family. My closet was primarily filled with hand-me-downs from my older brothers or meticulously sewn creations crafted by my mom's loving hands. Initially, I felt a pang of embarrassment, a longing for the trendy brands and styles my peers flaunted. I just wanted to fit in.

However, as I matured, my perspective totally shifted. I began to appreciate the unique charm and personal touch that these seemingly humble garments possessed. They were imbued with history, with the stories of my brothers and the love of my mother's tireless efforts. I learned to embrace their imperfections, to see the beauty in their individuality. While I may not have sported the latest fashions, I discovered a far more valuable lesson: I could express my own unique style, a style built on authenticity and personal flair, something that money simply couldn't buy.

From trying to lasso aggressive dogs from a tree to proudly wearing hand-me-downs, growing up on a farm truly taught me some of life's most valuable lessons. It wasn't always easy, but it certainly made me who I am today.

It wasn't about what I lacked, but about what I had: resilience, resourcefulness, and a deep

appreciation for the simple joys and the endur-
ing power of family and friendship—even if that
friend had four legs and a wagging tail. And
that, I realized, was wealth that no store-bought
wardrobe could ever provide.

From Kitchen Nails to Pin Cushion Universes

Living in our sprawling farmhouse, which we shared with my grandparents, I was definitely one of those kids whose mind just couldn't help but question the everyday. Our kitchen, with its charmingly old-fashioned vibe, wasn't just a place for meals; it was a treasure trove of everyday mysteries just waiting to be unraveled. And one particular enigma captured my attention more than most: the makeshift towel hanger. It wasn't anything fancy, just a series of sturdy nails hammered into the wall. And it was one of those nails that really triggered a thought experiment that occupied my little mind for a good half-hour.

I found myself staring at it, lost in a childish contemplation of the invisible forces that held the world together. What, I wondered, would happen if the atoms that comprised the wall around the nail suddenly decided to go their separate ways? Would the nail, deprived of its support, simply fall out? My imagination ran wild from there. If the atoms in the wall could spontaneously disassociate, what was stopping the floor from doing the same? The image of plummeting through the farmhouse floor, crashing through to the unknown depths below, filled my mind. It was a terrifying and exhilarating thought.

For what felt like an eternity, I stood there, transfixed. The nail remained firmly embedded, the wall steadfast, and the floor solid beneath my feet. The universe, it seemed, was holding its atomic act together, despite my theoretical musing.

Looking back, it was a wonderfully innocent and profoundly insightful experience. That simple nail, hammered into the wall of our old farmhouse kitchen, sparked a curiosity about the fundamental building blocks of reality. It was a reminder that even as a child, I was fascinated by the intricate dance of atoms, the unseen forces that maintain order in a seemingly chaotic universe. More than anything, it instilled a deep sense of wonder. It was amazing that everything, from the walls around me to the very floor I stood on, was held together by these invisible bonds. And it was even more amazing that I didn't have to worry about suddenly crashing through the floor, a realization that brought a quiet, satisfied smile to my young face. That kitchen nail, and the atomic anxieties it inspired, became a lasting testament to the insatiable curiosity and boundless imagination of a child exploring the mysteries of the world.

But my curiosity didn't stop with analyzing nails in the kitchen. My mom was an avid seamstress and had her sewing machine and accessories permanently set up at her sewing station. I found myself observing the pincushion full of straight pins with their small heads beside her sewing machine. To me, it looked like a mini universe. I wondered if this represented us in a small world, and that perhaps there was someone like me looking down on us, and that we were just a small part of something much, much bigger.

From kitchen nails to pin cushion universes, my childhood was a constant exploration of the everyday wonders around me. It just goes to show that you don't need a telescope or a microscope to unlock the mysteries of existence; sometimes, all it takes is a questioning mind and a bit of imagination, even in the most ordinary of settings. And honestly, I still look at nails and pincushions a little differently to this day.

Childhood Memories Etched in Time

Of course, that kitchen nail wasn't the only foundation I was building upon as a kid. Childhood, for me, was a tapestry woven with simple joys and enduring comforts. Two threads stand out vividly: the endless possibilities of a sandbox and the warmth of a kitchen filled with the aroma of cinnamon and chocolate.

My sandbox wasn't just a box filled with sand; it was a portal to another dimension. A dimension where I was the architect, the engineer, the creator of my own little world. Hours melted away as I meticulously crafted houses, castles, roads, and entire cities from the pliable sand. Armed with imagination and a collection of toys, I transformed a humble patch of sand into a landscape brimming with life. Water became rivers and moats, sticks and leaves served as intricate details, and each creation felt like a grand masterpiece. The sense of pride I felt in these ephemeral constructions was immense, a feeling of accomplishment that only a child's boundless imagination could conjure. Of course, the magic wasn't always perfect. The rude awakening of discovering a feline contribution after a night's slumber was a frequent reminder that my domain wasn't entirely my own!

Beyond the sandbox, the heart of our bustling family home was undoubtedly the kitchen. We were a large family, but my mother always ensured we were well-fed. Her dedication to homemade meals, even crafting pizzas from scratch, was a testament to her resourcefulness and love. And while she excelled at preparing hearty meat-and-potatoes dinners, one memory stands particularly clear: a large pot of steaming hot chocolate bubbling on the stove, accompanied by a tray of freshly baked cinnamon rolls. This wasn't dessert; this was a meal in itself, a blanket of warmth and sweetness on a chilly evening.

To this day, that combination of hot chocolate and cinnamon rolls remains my ultimate comfort food. Whenever I'm feeling under the weather, physically or emotionally, that familiar taste pulls me back to the coziness of my childhood kitchen. The sandbox might be long gone, and the house where I grew up might be filled with different families, but the memories remain. They are the sturdy foundations upon which I've built my life, reminding me of the simple joys that truly matter: the power of imagination, the comfort of a warm meal, and the enduring love of family. These are the threads that continue to weave through the tapestry of my life, keeping me grounded and grateful for the simple beauty of a well-lived childhood.

From Childish Confusion to Cosmic Clarity

Just as I wrestled with the invisible forces holding a nail in the wall and found comfort in the tangible creations of sand and food, I also grappled with concepts far beyond my immediate understanding in other areas of my life. I vividly remember learning the powerful opening verse of the Bible: "In the beginning, God created the heavens and the earth" (Genesis 1:1 KJV). While the notion of creation resonated, the word "heavens" presented a perplexing enigma. My young mind struggled to define its boundaries, imagining it stretching far beyond the vast blue expanse I could see.

For a child, the "heavens" felt like something just out of reach, a distant and mysterious realm. But as I grew, my perception of the "heavens" began to evolve. I realized that the word encompassed the sky above us, the twinkling stars, the swirling planets, and the endless tapestry of galaxies that stretched across the cosmos.

Suddenly, Genesis 1:1 took on a new, even more profound meaning. It wasn't just about creating something "out there" beyond our visual range. It was a declaration that God created everything—the Earth beneath our feet and the entirety of the universe above. From the smallest atom to the largest star cluster, all of existence was a product of His divine creation.

Understanding the breadth of "heavens" in this context clarifies the true meaning of the verse. It is not just about the physical act of creation, but about the absolute and complete origin of all things. God didn't modify or repurpose existing material; He brought everything into being from nothing. This verse acts as the bedrock of faith, the very foundation upon which all other theological concepts are built. Thinking of it like starting a new project helps solidify this understanding. Before you can build anything, you need a strong and stable foundation. In the same way, Genesis 1:1 lays the groundwork for comprehending God's power, His purpose, and His relationship with humanity. So, the next time you gaze up at the night sky, remember the simple yet profound truth of Genesis 1:1: "In the beginning, God created the heavens and the earth."

Looking back, my childhood was a fascinating journey of discovering foundations, each, in its own way, teaching me about the incredible order and wonder of the world, building blocks for understanding the grand tapestry of life itself.

Summer Nights and Soaring Rockets

I used to spend countless summer nights totally lost in the expanse of the night sky. A blanket was my trusty companion as I lay there, trying to figure out the secrets whispered by the twinkling stars. Sneaking out of the house at night was my specialty back then. I would just plop down on the grass, utterly mesmerized by the sheer volume of celestial bodies, each one a tiny point of light hinting at infinite possibilities. The silence of the night made the magic even louder, making me feel both tiny and profoundly connected to something way grander than myself. Gazing at the Moon, I would wonder if, with a powerful enough telescope, I could actually glimpse the footprints left by Neil Armstrong and Buzz Aldrin.

One summer night, driven by pure childlike curiosity, I decided to count the stars. The futility of the task became clear very fast. There were just too many! But instead of getting frustrated, I just succumbed to the beauty of the boundless night. The vastness was overwhelming, but in a weird way, it was also strangely comforting. A wave of peace washed over me as I just surrendered to the moment, appreciating the simple, profound beauty of the starlit sky. It was a solid reminder that sometimes, the best thing we can do is just be present and appreciate the wonder around us, without overthinking it.

Even before many of those epic star-gazing nights, I was totally captivated by rockets, the whole Apollo Moon program, and the truly awe-inspiring achievements of its astronauts. Their moonwalks weren't just events; they were like seeds planted deep in my young imagination. Growing up without a television, our family would gather at our neighbor's house to witness these historic moments. I remember lying on the floor, just as I did beneath the stars, filled with this incredible sense of awe and wonder. This same feeling fueled my attempts to construct my own makeshift rockets in the backyard, dreams taking shape in cardboard and a whole lot of hope.

Later, as a teen, I had a job as an auction runner, which was perfect for earning my own pocket money. The auctions, held in the evening on the first Thursday and last Saturday of the month, were a treasure trove of possibilities. My job was straightforward: carrying items to winning bidders and handling the financial exchange. But the real perk was the sneak peek I got before the auction even started. I would meticulously examine the contents of each box, strategizing which ones to bid on. Sometimes, I would unearth hidden gems—antique bottles, a vintage toolset. Through careful saving, I managed to accumulate enough to purchase something truly special: model rockets.

Building and launching those rockets was an absolute blast. It is like reliving those enchanting summer nights spent beneath the stars, remembering the first steps taken on the lunar surface by Neil and Buzz. One of my prized possessions was a Mercury Redstone rocket, a real testament to the pioneering spirit that ultimately paved the way for the Apollo program. Building these models felt, in a way, like mirroring the Apollo mission itself. The entire process, from meticulously assembling each component to watching my creation ascend into the sky, is exhilarating and deeply rewarding. It was, without a doubt, worth every penny I saved. It was a piece of the sky I could hold in my hands, a childhood dream finally taking flight.

There's a thrill in watching a rocket pierce the sky, a miniature spectacle of engineering and controlled explosion. For me, model rocketry was way more than just a hobby; it was a gateway to understanding physics, aerodynamics, and the satisfying crunch of a successful launch, flight, and recovery. But like any pursuit, the quest for bigger and better can lead to unforeseen challenges, especially when you start pushing the boundaries of "recommended" specifications.

Imagine having a sleek, well-designed rocket, practically begging to explore higher altitudes and reach greater speeds. The natural inclination is to upgrade the engine—to just slap on a larger, more powerful motor and watch it soar! It is an enticing idea, fueled by that raw desire to push the limits of what's possible.

However, this pursuit of power isn't without its risks. In the world of model rockets, it is not just about getting the biggest engine you can find. It is about finding that perfect balance between power and control, a delicate dance between thrust and stability.

The engines in these models, typically solid-propellant motors with no moving parts, come in various sizes, often designated by letters from A to D (and beyond, if you get serious). Each rocket model is designed with specific engine sizes in mind, recommendations carefully crafted by the manufacturer to ensure optimal performance and safe recovery.

But what happens when that inquisitive mind takes over? What happens when the urge to experiment outweighs the instructions? This is where I found myself modifying the motor compartments of my rockets, originally intended for B motors, to accommodate the significantly more powerful D motors. The allure of a higher altitude, a faster flight, was just too strong to resist.

My immediate thought was, "I need more parachute protection!" The higher and faster the rocket flew, the greater the potential for damage during descent. The higher altitude meant more time for the heat generated during launch to affect the parachute. However, my solution was based more on intuition than on rigorous calculation. I just guessed at the amount of additional protection needed, which, in hindsight, was a potentially disastrous approach in a pursuit where precision is paramount.

I quickly learned that understanding how an extra kick of power affects everything from how stable your launch is to whether your parachute even survives the ride down is the secret sauce. Without that foresight, all your big dreams just might come back to Earth with a less than graceful thud.

Looking back, two incidents from my model rocket days really stand out, proving that proper planning and, crucially, building in some wiggle room (what the pros call "design margins") are absolutely essential. And this isn't just for multi-billion-dollar NASA missions; it applies even to your backyard projects.

First up, there was the "super-powered three-engine" attempt. I decided to significantly upgrade the engines on my three-engine rocket. The launch? Breathtaking. Flames roared. That rocket shot upward with amazing speed. For a glorious few seconds, I was convinced I was a rocketry genius. But my moment of triumph was short-lived. When the parachute deployed, or rather, attempted to deploy, all I saw was a sad, shriveled mass of melted plastic. Turns out, those beefed-up engines generated so much heat that the parachute, which was designed for smaller engines, just couldn't handle the thermal stress. The rocket was mostly intact but definitely sported some new dents upon its unceremonious return to Earth.

Then there was the infamous "pencil rocket" incident. I was intrigued by its sleek design and the promise of ridiculous altitude, so naturally, I slapped a significantly larger engine on this single-stage beauty. The result was genuinely spectacular. It shot up so fast I practically strained my neck trying to track its trajectory. The problem, as always, arrived on the way down. I completely ignored the fact that it had gone way higher and faster than ever before. So, I failed to adjust the parachute size or even bother adding a vent hole in the center of the parachute to relieve pressure. Consequences? The rocket drifted... and drifted... and drifted. We watched its trajectory as it was floating back to the ground, but we couldn't find it in the woods. I could have been stuck in a treetop.

Farewell, sleek, high-flying friend. You were too beautiful for this world (and my lack of planning).

These experiences, as frustrating as they were at the time, hammered home some invaluable lessons about giving your systems proper "margins." In my youthful zeal, I was *only* focused on the "go-fast" part of rocketry, completely forgetting the crucial details needed to bring it home safely. Adding bigger engines isn't inherently a bad idea, but it absolutely cranks up the risk if you don't think about how your other systems (like, say, the parachute) are going to cope.

From Backyard Mayhem to NASA Precision

These early experiments, with their fierce successes and equally ferocious failures, truly laid the groundwork for a deeper understanding that kept developing during my time at NASA. Imagine my model rocket hiccups, but with far, far higher stakes. At NASA, a failure wasn't just about a bent rocket or a lost afternoon; it could mean squandered scientific data, compromised mission objectives, or, in the worst-case scenario, serious injuries or even lives lost. That's why meticulous design and those all-important design margins aren't just good ideas; they're non-negotiable lifelines.

Design margins are essentially your engineering safety nets, built-in redundancies that allow you to handle unexpected performance issues or extreme environmental conditions. They're there because, let's face it, engineering isn't always a perfectly predictable science. They ensure your system can handle stresses and strains way beyond what you normally expect. My melted model rocket parachute? That was a textbook example of needing a thermal margin. The drifting pencil rocket? That screamed for aerodynamic margins to handle those higher speeds and altitudes.

Now, if you want a master class in "what not to do," let me tell you about the "dud rocket engine" incident. We've all been there: that moment a genius idea pops into your head, only to explode in your face (sometimes literally!). Think of that time you tried to supercharge your bike, only to spend the next hour wrestling with a perpetually skipping chain. Now, multiply that frustration by ten, add a healthy dose of uncontrolled fire, and you'll get a pretty good sense of my ill-fated experiment.

I was a kid with a problem: a growing collection of useless rocket engines, relics from past hobby adventures that had stubbornly refused to ignite. Fed up with the engines not working, I hatched a plan. Why not light them up? It seemed logical enough at the time, a fiery send-off for these underperforming projectiles. What could possibly go wrong? (Spoiler alert: everything.)

My staging ground was a large gravel patch between the garage and barn. My fuel of choice? A cardboard container filled with a volatile mix of gas and oil salvaged from a discarded cardboard car oil container. Yeah, I know. Retrospect is a powerful, cringe-inducing thing. Let's just say it wasn't my brightest moment.

The second I applied a flame to those engines, the situation spiraled into absolute chaos. My vision of a controlled, contained ignition was replaced by a chaotic ballet of fire and near-catastrophe. The first engine roared to life, spitting flames and sparks, before launching itself a staggering 400 yards into a nearby field. Then the second one ignited. My heart practically leaped into my throat as I watched it hurtle dangerously close to the open doors of the barn, which, by the way, housed a mountain of dry, highly flammable hay. I was inches away from igniting a massive inferno, endangering myself and my little brother. Thankfully, the rogue engine narrowly missed its target.

Panic took over. I knew I had to stop this madness. In a desperate, split-second attempt to extinguish the flames, I kicked the oil can, sloshing the fuel and fire in every direction. Unfortunately (and adding insult to injury), some of it landed on my younger brother's jacket, setting him ablaze. He panicked too, and started running toward the house, intent on plunging into our basement's water trough—a large cement trough filled with water from a natural spring.

Knowing that bringing the fire inside the house would likely get me in trouble and could escalate the situation into a full-blown house fire, I tackled him to the ground. Not exactly a textbook maneuver, but it worked. The flames were smothered. It was a knee-jerk reaction, a real-life "stop-drop-and-roll" performed without a second thought. Sometimes, you just get lucky.

The immediate crisis averted, I was left with the embarrassment of my utter failure. My parents remained blissfully unaware of the near-apocalyptic chaos I had unleashed until much, much later.

The lesson, learned the hard way, was crystal clear: never, ever play with fire, especially when it involves dud rocket engines, volatile fuel mixtures, and a complete lack of common sense. My rocket engine fiasco remains a vivid reminder that even the most seemingly ingenious ideas can backfire spectacularly, leaving you covered in soot, shame, and a profound understanding of the potential consequences of reckless experimentation. And most importantly, it taught me the value of a good fire extinguisher.

So, go ahead, dream big and reach for the stars. Ambition is awesome! But remember to do your homework first. Understanding the impact of increased power on every aspect of your rocket's performance—from launch stability to parachute deployment to not setting your brother on fire—is the key to transforming that ambition into a truly successful flight. The sky's the limit, but a little planning can help you get there safely. And maybe add a few extra fire extinguishers, just in case.

Living on the Edge

My childhood, a rich tapestry woven with bold attempts and sweet comforts, still yields a trove of fond memories that linger vividly in my mind.

We all have those moments in life that leave us breathless, perched on the edge of possibility (or maybe a little bit of danger). Those times when you look back and wonder, "How did I even survive that?!" They're the unexpected twists and turns that make life an exhilarating, if sometimes bewildering, ride. For me, those wild moments were often fueled by the excitement of the annual County Fair.

Growing up, the County Fair was the event of the year. I would spend days soaking in the sights, sounds, and smells—the aroma of cotton candy, the kaleidoscope of exhibits, and the roaring engines of the demolition derby, which was a particular favorite. The sheer chaos and destruction, the cheering crowds; everything about it was captivating. And then there were the car races on the dirt track, a blur of speed and skill as drivers drifted around corners, vying for the lead. But nothing compared to the Joey Chitwood show, a spectacle of precision driving where he'd execute impossible maneuvers, including driving on two wheels!

For a kid, witnessing such daring feats ignited the imagination. Naturally, these adrenaline-pumping experiences translated into backyard adventures. Armed with this newfound inspiration, I decided to recreate the thrill of the racetrack on our farm.

My first attempt involved the family tractor and a secluded field. I envisioned myself as a champion racer, navigating the uneven terrain with daring precision. Each lap was faster, more ambitious, until the inevitable happened. The tractor tilted precariously on two wheels during a sharp turn, threatening to flip. Luckily, I managed to regain control, shaken but unharmed. My parents remained blissfully unaware of my little escapade, at least until years later.

Undeterred, but realizing the tractor wasn't the ideal racing machine, I switched to a more suitable vehicle: the family car. This time, I chose a flatter field, though it was unfortunately visible from the house. The lower center of gravity made me feel like I was truly flying, the adrenaline coursing through my veins as I pushed the car to its (and my) limits. The thrill was short-lived, however, as my dad quickly put an end to my impromptu racing career after spotting the tire tracks in the field.

Beyond the excitement and near-mishaps, these experiences taught me valuable lessons. More than the thrill of pretending to be a race car driver, I realized the importance of responsibility and the impact of our actions.

> As young people, we are incredibly impressionable, constantly observing and learning from the adults around us. Parents, teachers, and

*even slightly older friends shape our perspec-
tives and behaviors.*

Then there was the time I tried to jump a ramp on my bicycle. I recall the sheer audacity of my bicycle ramp adventure, where hours of effort went into constructing crude launching and landing zones. The thrill of hitting that first ramp, soaring through the air, and for a fleeting second believing I had conquered gravity, was immense—a dream that crashed just as quickly when my wheels met the unforgiving ground a foot short, sending me tumbling in a chaotic mix of metal and limbs.

Yet, the spirit of invention wasn't deterred; by thirteen, my trusty Honda 90 motorcycle, typically a fair-weather companion, became an unlikely winter warrior. I used a downhill ski strapped to the front wheel, transforming a motorcycle into a silent, exhilarating glide across deep snowdrifts, a testament to youthful ingenuity and the thrill of pushing boundaries. But amidst these daring escapades, there was always the comforting anchor of my great-grandmother's house, a mile-long trek across an open field that always promised a sweet reward. As I approached, the air itself seemed to sweeten, and pushing open her front door confirmed the anticipation: the unmistakable, comforting aroma of her perfect sugar cookies, a warm, delicious embrace that filled every corner of her cozy home, serving as the perfect, grounding counterpoint to a boy's daring adventures.

My time on the farm, even beyond the misguided racing attempts, ingrained in me the value of responsibility. Caring for animals and crops demanded a commitment to their needs, often before my own. This instilled a sense of accountability and taught me the importance of discipline, integrity, and hard work. Waking up early to tend to the fields, despite my exhaustion, reinforced the importance of sticking to my commitments.

Ultimately, the County Fair, the thrill of the races, and the resulting farm adventures served as a unique classroom. Beyond the adrenaline rush and the near disasters, they provided invaluable lessons in responsibility, the power of influence, and the profound impact even seemingly small actions can have. They were a chaotic, exciting, and ultimately formative part of growing up. And, thankfully, I survived to tell the tale.

Chapter Two

School Experiences—The Early Years

S chool. For some, it is a place of academic achievement, soaring grades, and a clear path to success. For
others, like me, the early years were a mix of entrepreneurial spirit, social navigation, and the slow realization
of education's true value. Looking back, these formative
experiences, from the seemingly insignificant to the profoundly impactful, shaped me in ways I later appreciated.

My journey into business started early, in the first grade
to be exact. Fueled by a boundless ambition and a desire
to earn (a whopping) two pennies per riddle, I launched a
lunchtime riddle-selling enterprise. I spent hours crafting clever jokes and compiling a menu categorized by
animals, food, and general humor. To my surprise, it actually worked! My classmates enjoyed my creations, making
me, for a brief period, the riddle king of the first grade.

It is amusing to think about the seriousness with which I approached this venture, but it taught me a valuable lesson: everyone starts somewhere.

While my entrepreneurial spirit thrived, my academic drive lagged. I found much of the curriculum in those early years "boring and pointless," leading me to under-achieve and squander my potential. It wasn't until much later that I grasped the crucial role that education plays in future success. A solid foundation is essential for ex-celling in any field, and I now understand that it is never too late to change your attitude toward learning.

One significant lesson came from my third-grade teacher, Mrs. Johnson (not her real name). She instilled in us the importance of voting as a way to have a voice in society. However, she also emphasized the privacy of voting and the need for thoughtful consideration before sharing our choices. It was a pretty big concept for a bunch of third graders, but it stuck with me.

Fourth grade brought with it a monumental struggle: math. I was on the verge of failing, a fact that only dawned on me years later. Thankfully, my mother stepped in. De-spite only having an eighth-grade education herself, she dedicated countless hours to helping me with my home-work at the kitchen table. Her patience and dedication, even though I questioned the need, helped me improve, albeit gradually. I will always be grateful for her unwaver-ing support, even when the subject matter presented a challenge for both of us. Honestly, I think we both learned a lot during those sessions!

Fifth grade marked a shift toward the social sphere. While homework and tests were still present, the focus shifted to friendships, gossip during recess, and navigating the complex social dynamics of elementary school cliques. This year was a formative period for building relationships and understanding the intricate web of peer interactions. Who knew playground politics could be so intense?

Interestingly, my fifth-grade teacher recognized leadership qualities in me before I even recognized them myself. He would often assign me special tasks, like leading group projects or writing class skits. These experiences, though initially nerve-wracking, provided opportunities to develop my skills and build confidence. Looking back, I realize the profound impact he had on shaping my leadership potential, gently pushing me outside my comfort zone, and offering words of encouragement.

My early school years were a patchwork quilt of experiences: a budding business venture, a gradual understanding of educational values, a lesson in civic responsibility, a mother's unwavering support, and the discovery of hidden leadership potential. These experiences, though seemingly disparate at the time, have woven together to help form the foundation of who I am today.

How Childhood Chores Shaped Me

Growing up, I wasn't exactly destined for greatness, or so I thought. My true calling, at least in my mother's eyes, lay in the realm of household chores. Need an extra jar of pickles from the depths of our basement? I was the go-to guy. A forgotten bottle of hot sauce to add a kick to dinner? I knew the exact shelf in our dimly lit storage area. Looking back, it wasn't glamorous, but it unexpectedly shaped me.

I remember countless expeditions down those creaky stairs or the storage room in pursuit of some crucial grocery item. It became so routine that even now, I could probably locate the dill relish with my eyes closed. While I might have occasionally grumbled (mostly internally, because, you know, mom), those little errands instilled a sense of responsibility, even if it was just the responsibility of being the designated pickle-getter.

But the most interesting byproduct of my constant chore-doing was a brief, almost comical, foray into forced optimism. Somewhere along the line, I developed the habit of saying "could have been worse" whenever something mildly annoying happened. Stub my toe? Could have broken it! Hit my fingernail with a hammer? Could have split it clean off! It was my bizarre way of coping with the everyday frustrations of life, a self-inflicted attempt to find the silver lining.

I'm not entirely sure where this "could have been worse" mantra originated, but I suspect it was a subconscious defense mechanism, born out of a childhood where small chores were a regular occurrence. It was a quirky way to reframe minor inconveniences. Sadly, or perhaps thankfully, this habit didn't survive in my teenage years. Imagining a brooding teenager constantly muttering "could have been worse" after a bad hair day just doesn't quite fit the angsty aesthetic.

Beyond the chores and the fleeting optimism, I also vividly recall observing my mom's own coping mechanisms. When she was feeling overwhelmed by the pressures of daily life, she'd retreat to the woods across the fields. It was her sanctuary, a place to escape, even if just for a little while.

> *This experience taught me the profound importance of finding your own space, your own way to de-stress and recharge your batteries.*

So, while I might no longer be the designated pickle-getter, and the "could have been worse" attitude is but a distant memory, those childhood experiences, even the seemingly mundane ones, undeniably shaped the person I am today. They taught me responsibility, highlighted the value of a good coping mechanism (however short-lived), and impressed upon me the importance of finding solace when life gets overwhelming. And who knows, maybe one day I'll rediscover my inner optimist, one "could have been worse" at a time. Until then, I'll simply appreciate the lessons learned while navigating the dusty shelves of my childhood.

Faith, Fellowship, and a Coleman Stove on the Road

Family vacations often conjure up images of sun-drenched beaches, thrilling theme parks, or the effortless luxury of a cruise ship. But for my family, the annual summer pilgrimage took a different form—a cross-country adventure fueled by faith, fellowship, and a station wagon brimming with snacks: the annual church conference.

Forget meticulous itineraries and pre-booked excursions. This was a journey into the heart of community, a tapestry woven from shared beliefs and the open-armed generosity of strangers. Picture it: our station wagon, packed to the gills with kids, luggage, and the unspoken promise of a new adventure unfolding with every mile. The back seat, a coveted and constantly contested space, became a rotating battleground of limbs, each child (myself included!) mastering the art of contorting themselves into the most comfortable (or at least, the least uncomfortable) position.

Luxury hotels with room service were definitely out. We traded them for a unique experience in rustic charm and community spirit. Accommodation was a masterclass in grateful acceptance, as families bunked with host church members, their homes generously opened to weary travelers. Those lucky enough (and nimble enough to claim them!) snagged precious spots in one of the portable camping trailers that dotted the landscape. It was a profound testament to shared faith and a willingness to lend a helping hand, a beautiful display of kindness and generosity that always stuck with me.

But the unsung hero of our annual adventure, at least through the eyes of a child, was the humble Coleman camp stove. Forget gourmet meals prepared by a five-star chef. My mother, a true culinary magician, conjured up feasts on this portable marvel. The rhythmic pumping of the fuel tank, a necessary ritual to build the pressure to keep the flame alive, was almost a meditative process. The contented hiss of the gas, the aroma of food sizzling in the open air—these are the sensory memories that remain etched in my mind, forever linked to the warmth of family and the spirit of adventure.

Looking back, these weren't just vacations; they were profound experiences. They were about connection, community, and appreciating the simple joys of life. We learned to embrace the generosity of strangers, the importance of sharing what we had, and the unmatched deliciousness of a meal cooked over that trusty camp stove. I learned patience, adaptability, and the value of making memories in the most unconventional settings.

While they might not have been the idyllic family vacations plastered across glossy travel brochures, they were uniquely ours. They shaped me in ways I'm only beginning to fully understand. They served as a powerful reminder that the best adventures aren't always about reaching the destination, but about cherishing the journey itself, and the people you share it with. And maybe, just maybe, a little bit about that trusty Coleman stove, a silent symbol of resourcefulness, resilience, and unforgettable family memories. In a world that often prioritizes extravagance, these simple, faith-filled road trips taught me the true meaning of richness.

A Journey Rooted in Faith and Song

My faith journey didn't kick off with some big, dramatic "aha!" moment or by myself in a quiet room. Nope, it started right in the warm, cozy embrace of my family. Faith wasn't just a Sunday thing we pulled out and dusted off; it was woven right into the fabric of our everyday lives, a constant presence that shaped our values and guided our actions.

I can still picture those evenings, huddled up on the sofa with my mom, totally lost in the captivating world of Bible stories. Man, she had a knack for bringing those ancient stories to life, filling them with humor, wisdom, and a deep, deep sense of reverence. Those weren't just bedtime stories; they were like little seeds of faith, gently planted in our hearts, watered by her unwavering love, and nurtured in that special space we shared.

One memory really shines bright. I was a young boy, super eager to follow in my older brother's footsteps, who seemed to embody everything I aspired to be. But my mom, with her gentle wisdom, saw my eagerness and knew I needed my own journey. She reminded me that my relationship with Christ was just that—mine. I didn't need to wait for my brother; I could choose to embrace faith on my own terms. That subtle encouragement really stuck with me, a quiet "yes, your path is valid and uniquely yours." And so, at lucky thirteen, in the quiet solitude of my room, I got down on my knees and, with a heart full of hope, just asked Jesus into my heart.

My baptism? That memory is still so clear in my mind, like it was yesterday. I can still feel the cool water washing over me, that amazing sense of renewal and commitment filling my soul. But what really sticks with me is the image of my grandfather, standing tall and proud, his eyes just full of love and blessing as he said a simple, yet incredibly profound thing: "God bless you, Geoffrey." His words felt like a warm hug, a confirmation of the path I was choosing, and a testament to the awesome legacy of faith passed down through generations.

For our family, life truly revolved around our small church, a place that was our absolute favorite spot for community and shared belief. But the absolute heart and soul of our church experience, what really made it for us, was the music. And nope, no fancy instruments or choir robes there; it was all about the pure, heartfelt, a cappella harmonies that didn't just fill a small country church but seemed to flow, expanding to resonate deep within our very souls. What made it even more profound was how the hardwood floors and wooden walls of the sanctuary seemed to come alive, resonating each note in this majestic, all-encompassing flow that just enveloped everyone. Singing in four-part harmony without instruments really taught us to listen—not just to the melody, but to the individual voices around us. We learned to blend, to create this unified sound of praise that just soared to the heavens. That act of communal singing, of putting your own voice aside for the sake of a harmonious whole, became a powerful metaphor for how we tried to live our lives, striving for unity and love within our family, our church, and our community.

Growing up attending a small rural church

Music just has this incredible power to stir your soul, bring back memories, and lead you on totally unexpected journeys. For me, gospel music has been this constant companion on my spiritual path, a journey super intertwined with my dad's influence. He was a man of diverse passions—he *loved* livestock auctions—and funny enough, he unwittingly kicked off my lifelong love affair with gospel music.

It was at one of these auctions that my dad first stumbled upon the Cathedral Quartet, where they were performing a few of their songs during a break from the auction. He was totally captivated by their performance, and then he made sure they got opportunities to sing at different venues. They quickly became my favorite quartet, their harmonies a comforting hug, and their message a source of pure inspiration. Among them, George Younce, the bass singer, held a super special place. His rich, resonant bass voice struck a chord deep down, and he ended up being a real guiding light as my own bass voice began to develop. Years have flown by, and sadly, two key members aren't with us anymore, but the Cathedral Quartet is still my go-to for comfort and upliftment. Their music still hits with the same power and grace it always has.

Beyond the quartet, my dad also introduced us to the powerful story of Tony Fontane. Fontane, a tenor singer whose life was totally flipped after a near-fatal car accident, dedicated his renewed life to singing gospel music. Watching the movie about his journey and hearing his incredible voice was especially moving, especially since my own dad had a beautiful tenor voice himself, which he generously used to bless others through song. Fontane's testimony of faith and the power of music really left a mark.

So, these memories—all woven together with faith, family, and the sweet melodies of gospel music—paint a pretty vibrant picture of my journey with Christ. It started with those bedtime stories and a gentle nudge from Mom, blossoming into a lifelong relationship nurtured by the unwavering support of loved ones and the uplifting power of music. From the deep bass tones of George Younce to the soaring tenor of Tony Fontane, gospel music has truly been the soundtrack to my spiritual growth. And it all kicked off with my dad, a livestock auction, and his deep appreciation for how music can touch your heart and uplift your soul.

My journey with Christ is a mix of family, faith, and song. It began with those bedtime stories and gentle encouragement, blossomed with personal conviction, and has continued to be nourished by the enduring power of community and the transformative beauty of music. It is a journey that keeps unfolding, shaped by those seeds of faith planted so long ago, nurtured by love, and guided by the unwavering grace of God.

Chapter Three

Extracurricular Activities

My Musical Journey Through School

My love affair with instrumental music began in seventh grade, a time when clarinet reeds and sheet music seemed like a foreign language. I decided to dive headfirst into the school band, completely oblivious to the nuances of instruments or musical notation. Following my band director's advice, I initially chose the E-flat horn, and later, the French horn.

The initial days were, to put it mildly, challenging. Instead of melodious tones, my instrument primarily emitted squeaks and squawks! But perseverance, fueled by daily practice and dedicated music lessons, gradually transformed the cacophony into proficiency. I learned to coax the bright, bold sound from the E-flat horn, a sound that added a special flair to our performances. Later, I mastered the warm, elegant tones of the French horn.

Being part of the school band was so much more than just learning to play an instrument. It was about developing my musical skills, fostering camaraderie with fellow music lovers, and discovering a passion that would shape my high school years. The culmination of all those hours of practice came in the form of being selected to play the French horn in the All-County Band from tenth through twelfth grades, and even the State FFA Band in twelfth grade. The rigorous audition process was daunting, but the reward of performing alongside some of the best musicians in the state was truly surreal. Traveling and showcasing my skills at various events felt like icing on the cake!

The All-County Band experience was more than just about the music; it was about the connections I forged. I remember sitting next to an exceptionally talented French horn player, whose sharp wit and humor instantly drew me in. We bonded over our shared passion for music, and she even offered invaluable tips on refining my technique, helping me mellow out the often-brassy sound I produced. This unexpected friendship made the entire experience even more enjoyable, highlighting how a simple seating arrangement could lead to such lasting connections.

Friendship, back then, meant shared experiences. So, when my new friend invited me to attend an opera with her family, I jumped at the opportunity. The music was beautiful, the costumes extravagant, and the atmosphere elegant. Sharing that special moment with my friend and her family made me feel grown-up and sophisticated. While the intensity of rehearsals and performances brought us together, the end of the All-County Band practice and concert led to a disappointing loss of contact. Everyone went their separate ways; a stark reminder of how quickly life can change.

My dedication to mastering the French horn led to some unconventional experimentation. I experimented with the placement of the mouthpiece, alternating between the sides of my mouth until I could control the tone effectively from either side. I also practiced coordinating different rhythms with my hands and legs simultaneously, training my mind to multitask and enabling me to potentially play two parts of a musical score. It required practice and dedicated guidance from my band instructor, but it was a rewarding challenge.

This experimentation led me to the outlandish idea of playing two horns at the same time. To my surprise, I was actually able to play a simple song simultaneously on both horns! It was a fun and rewarding challenge that pushed me beyond my comfort zone.

While the French horn held a special place in my heart, I recognized its limited application in the marching band. So, I also played the E-flat horn and trumpet in the school marching band and joined the school jazz band playing the trombone. The energy and teamwork of the marching band performances were electrifying, while the smooth, soulful tunes of the jazz band allowed me to explore improvisation and express myself in a different way. One unforgettable moment was when the jazz band director started a song at a concert, then simply walked off stage, leaving us to improvise and groove on our own. These experiences added a layer of excitement and fulfillment to my high school years.

My parents, convinced of my deep bass voice, strongly encouraged (read made me) me to join both the school chorus and the community choral society. Initially hesitant, I decided to give it a try. I dedicated myself to the chorus and even took vocal lessons. My father also formed a quartet, with my brothers and him singing tenor, lead, and baritone, while I anchored the ensemble with my bass. I was initially resistant, but through their persistent encouragement and my diligent efforts, I developed a genuine appreciation for performing and singing in front of an audience.

This journey underscores a valuable lesson: sometimes, fulfilling our parents' wishes can lead to unexpectedly rewarding experiences, leading to the discovery of hidden talents and passions we never knew we possessed. From awkward squeaks on the E-flat horn to performing on stages across the state, my musical journey through school shaped me in ways I couldn't have imagined. It provided the needed distraction from my lackluster academic per-

formance. It was something I could identify with and own, independent of my 7th-grade teacher pronouncing my name incorrectly or being told I wouldn't amount to much.

Balancing Act: My Pennsylvania Farm, My Sports Dreams

Growing up in rural Pennsylvania, sports weren't just a pastime for me; they were a full-blown passion, fueled by the echoes of cheering fans all the way from Pittsburgh. Our local school might not have had the resources for a powerhouse football team, but the spirit of the Steelers, Pirates, and Penguins absolutely lived on through soccer, baseball, and basketball. I remember being totally inspired by the legendary Pelé and finding myself drawn to the soccer pitch. And with a super-talented cousin who was three years ahead of me in high school as my personal benchmark, I embraced the role of goalie in just my second year of playing soccer.

But life on the farm definitely threw a unique curveball into this whole sports dream. Balancing soccer practice and games with the constant demands of farm work required a level of discipline that, honestly, seems almost unbelievable to me now. My mornings kicked off before dawn with chores, followed by a day at school, and then practice or games, and finally, heading straight back to the farm for even more work. It was a seriously demanding schedule, but the lessons I learned from navigating that tricky balance stuck with me, shaping how I approach pretty much all of life's challenges today.

One particularly memorable part of this balancing act was my commute home from soccer practice. Since I didn't have a driver's license yet, I would literally jog the five miles through the beautiful, rolling hills in that part of Pennsylvania. At first, it was a grueling task, a real workout that left me breathless. But somehow, it slowly transformed into this routine, a surprisingly peaceful escape. Feeling the breeze on my face, knowing I was getting a workout *and* being productive by getting myself home, was surprisingly freeing. Of course, there were days when the lure of a car ride was super strong, especially if it was raining or I was just plain exhausted. But looking back, I genuinely cherish those jogs as some of the most tranquil moments of my teenage years. And let me tell you, when I finally earned that driver's license, that car felt like the ultimate reward!

Inspired by my older brother's involvement, I also decided to give basketball a shot. We even set up a makeshift basketball court right there in our barn—it was pretty cool! However, as the demands of both basketball and farm work started to intensify, trying to maintain that delicate balance proved pretty much impossible. Evening practices and games began to seriously eat into the time I needed for my farm duties. It wasn't just about being exhausted; it was about feeling completely torn between two equally important worlds. Ultimately, my dad made the difficult decision to ask me to quit basketball. As heartbreaking as it was at the time, I knew deep down that my commitment to the farm had to come first. It was a tough pill to swallow.

The disappointment of quitting basketball was sadly compounded by the fact that my parents never actually attended any of my games, whether it was soccer or basketball. After putting in so much time and effort into practicing and competing, I really longed to see them on the sidelines, cheering me on. Their absence made me feel like they didn't really value my interests and accomplishments. Seeing other kids surrounded by their supportive parents only amplified that feeling of hurt and disconnect. It definitely affected my motivation and left me questioning whether my passions even mattered to them.

Despite all the challenges and disappointments along the way, those experiences of balancing sports and farm work instilled in me a profound sense of discipline, responsibility, and resilience. These lessons, forged in the fields and on the court, continue to guide me today, reminding me that even in the face of adversity, dedication and hard work can truly pave the way for a fulfilling life.

A Motorcycle Misadventure

Being sixteen was a trip. That heady mix of testosterone, a desperate craving for speed, and just enough freedom to get yourself into trouble? Logic definitely took a backseat to impulse back then. That was certainly my vibe when I swapped out my trusty Honda 350 for a more powerful 1972 Honda 750. It was a beautiful beast, no doubt, but in the early 70s, there was always something faster lurking around. My neighbor, for instance, flaunted his Kawasaki 900, the undisputed king of the quarter mile. And you know what that did to a competitive farm kid? It ignited a fire. I simply had to find a way to make my Honda 750 keep up.

Now, let's be clear: I wasn't an engineer. I wasn't even a proper mechanic. I was just a farm kid with a wrench and a whole lot of misguided confidence. My brilliant solution? Tinkering with the carburetor, the very heart of my Honda. I decided to modify the fuel jets to boost fuel flow, which, in my teenage brain, translated directly to "more power!" So, armed with a tiny 1/16-inch drill bit and a plan as solid as wet sand, I set to work.

The theory was simple enough: more fuel equals more power. The actual execution, however, left a lot to be desired. After a few nervous minutes, the deed was done. I drilled out those jets, ready for the torrent of speed. I fired up the engine, a tremor running through me as the revs climbed. I roared down the road, bracing myself for a surge of untamed power.

Instead, I got sputtering. At higher RPMs, the engine choked and hesitated, a far cry from the smooth, throaty growl I craved. Disappointment washed over me like a cold shower. It was painfully obvious: my impromptu "upgrade" had completely backfired. I clearly had messed with the delicate fuel-to-air ratio, throwing everything out of whack.

But a farm kid isn't easily deterred! Thinking on my feet (and probably fueled by the lingering scent of gasoline), I had another epiphany. If the engine was getting too much fuel, I reasoned, I just needed to introduce more air. The solution was suddenly blindingly obvious: ditch the air filter.

With the air filter unceremoniously removed, I fired the engine up again. This time, the difference was absolutely remarkable. The sputtering was gone, replaced by a raw, unrestrained roar that sent shivers down my spine. I blasted down the same stretch of road, and this time, it felt... different. The bike pulled harder, faster, with a newfound aggressiveness I hadn't dreamed of.

Incredibly, my modification seemed to have worked! My Honda 750, stripped of its air filter and sporting overly generous fuel jets, actually felt comparable to my neighbor's Kawasaki 900. We never formally raced, mostly because a healthy dose of pride (and the very real potential for mechanical catastrophe) kept us from a head-to-head showdown. But I knew, deep in my heart, that my modified Honda could hold its own.

Looking back now, I honestly cringe at the sheer recklessness and ignorance of my actions. I was playing with fire (quite literally) and could have easily trashed the engine or, worse, seriously jeopardized my safety. But in that moment, fueled by youthful arrogance and the pure thrill of the ride, I felt like I had somehow outsmarted Honda itself.

Those early experiences, combined with the endless freedom of growing up on a farm, really forged a unique and unforgettable childhood. Not every adventure unfolded quite so smoothly, though, a lesson that would be driven home in the years to come.

Chapter Four

High School—The Troubled Years

A Chorus of Chaos and Questionable Choices

Ah, adolescence. That bizarre and bewildering period of fluctuating hormones, questionable fashion choices, and an almost pathological need to question everything. For me, it wasn't just a phase; it was a full-blown rebellion concerto, a symphony of angst and awkwardness conducted with the reckless abandonment only a teenager can muster. Looking back, I was a walking, talking teenage cliché, and my questioning wasn't just about the meaning of life, but a deep-seated, almost primal resistance to anything resembling authority.

My dad, a man of unwavering good intentions, bore the brunt of this rebellion. I vividly recall a crisp autumn morning when he simply asked me to check the tire pressure before driving to school. My response? A sarcastic barb dripping with teenage disdain. Cue the internal cringe that echoes through the years. Yeah, not my finest moment, Dad.

Then there was the singing. Oh, the relentless, inescapable singing! Blessed (or cursed) with a decent bass voice, I was required to participate in every musical ensemble within a five-mile radius. Church choirs, school choirs, community choirs, even a family quartet! My vocal cords felt perpetually enslaved. I even feigned laryngitis once to escape a concert, a ruse my dad, especially enamored with the family quartet and choral society, saw right through. In my teenage mind, this only fueled my resentment toward him and the whole singing charade. Sorry, Dad! I was young and... tuneless.

But my rebellion wasn't just about sullenness and eye-rolling. It involved some truly questionable decisions that could only be excused by the potent cocktail of teenage boredom and invincible ignorance. Being selected for All-County Chorus meant spending a few nights away from home, which, for a rebellious teenager, was a seriously dangerous combination. One evening, a fellow choir member and I took a shine to the director's Volkswagen Beetle. Let's just say hotwiring a car proved surprisingly easy. (For the love of all that is good, do not try this at home. Seriously.) We returned it, of course, but the sheer idiocy of it still makes me shake my head.

Speaking of questionable behavior during chorus rehearsals, there was this one particular song where a phrase ended with the word "chicken." My mischievous friend and I found it utterly hilarious to tack on "pox" at the end every single time. The director's amusement level? Less than zero. I'm pretty sure he aged a decade in that rehearsal alone. Honestly, we thought we were comedy gold.

My commitment to absurdity even extended to singing competitions. I once graced the stage with a performance of "I wish I were an Oscar Meyer Weiner" in a baffling mix of German and English. I didn't win, but I certainly provided some entertainment, albeit of the bizarre variety. The judges' faces were a mixture of confusion and amusement, a reaction I now realize perfectly encapsulated my teenage years.

Then there was the time I volunteered to sing with the youth choir at a sister church. During introductions at each concert, I felt compelled to claim I was the son of the host family or the choir director. The logic? Lost to the annals of teenage brains, along with common sense and good judgment. Why? I have no idea. Probably just to shake things up.

Amidst all the chaos, my Aunt Vera, a beacon of sanity and grace, tried to guide me. She constantly reminded me that my voice was a gift from God and neglecting it would have consequences. She even penned me a letter about it a few months before she passed, reminding me to use gifts entrusted to us. Her wisdom was, frankly, a bit lost on my hormone-addled brain back then, but I appreciate it now.

Of course, no teenage rebellion is truly complete without a healthy dose of reckless driving. One winter, a blizzard canceled school for a whole week. The snowplows carved massive walls of snow along the roads, which I, naturally, interpreted as an invitation to test my driving prowess, specifically, my ability to spin my car 360 degrees without causing damage. I successfully executed the maneuver once before realizing the sheer idiocy of my actions. The near miss was a sobering reminder that even in a blizzard of teenage angst, reality bites. Hard.

Then there was my trusty Honda 350 motorcycle. Rain or shine, it was my freedom machine, a throbbing metaphor for my burgeoning independence. I even braved a 30-mile downpour on it after a choir competition. Good times... I think. Everything felt more epic on that bike.

Looking back, my teenage years were a wild, chaotic, and often ridiculous ride. It was a time of self-discovery, even if that discovery involved making a multitude of mistakes. And hey, at least I have some unbelievably funny (and slightly embarrassing) stories to tell now. They're discordant notes that, somehow, create a uniquely memorable melody. A melody that, despite its imperfections, is undeniably mine.

How a Teacher's Pronunciation Snafu Changed My High School Trajectory

Seventh grade is a minefield. Awkward growth spurts, questionable fashion choices, and for me, an unexpected battle over the pronunciation of my own name. My English teacher became convinced that my given name, Geoffrey, undeniably pronounced the same as "Jeffrey," should instead be uttered as "Goffrey." She even, with bewildering confidence, cited Geoffrey Chaucer as justification, seemingly oblivious to the ever-evolving landscape of language and pronunciation.

My father even took the fight to the school board. Yeah, you heard that right—the school board, but the "Goffrey" saga persisted. It was a daily drip of annoyance, a constant reminder of being misunderstood and a persistent itch I just couldn't scratch. Frustration mounted, big time.

7th Grade High School picture

Fueled by righteous anger only a pre-teen can truly muster, I discovered a loophole in her grading system. The teacher operated under a seemingly benevolent rule: complete all homework, and you wouldn't fail, regardless of your test scores. I saw this not as a safety net, but as a challenge. I meticulously completed every assignment (sometimes even included the right answers), a picture of diligent effort on the surface. But behind the facade, I deliberately bombed the tests. My objective was clear: pass the class with the bare minimum effort, a silent protest against the persistent mispronunciation of my name. I succeeded, earning a triumphant D-. Looking back, I admit I could have aimed higher, but the principle, fueled by pre-teen indignation, seemed far more important at the time.

However, the repercussions of this situation rippled far beyond that single English class. The frustration and resentment, left unchecked, festered and subtly poisoned my attitude throughout high school. I was so consumed by my own perceived injustice, so focused on the pronunciation of "Geoffrey" vs. "Goffrey," that I failed to recognize the impact I was having on those around me. Some of my teachers noticed, a few even tried to address my sour disposition, but I was too entrenched in my bitterness to truly listen. I was convinced I was right, and everyone else was wrong.

It wasn't until years later, reflecting on my tumultuous high school experience, that I truly understood the lesson I should have learned. The "Goffrey" incident wasn't just about a name; it was about the power of attitude and the critical importance of recognizing its impact on others. While teachers are there to support us, guide us, and help us navigate the often-turbulent waters of adolescence, our own behavior can either foster a positive learning environment or inadvertently drag everyone down.

That misguided battle over my name, and the resentment it fostered, inadvertently shaped a significant portion of my high school years. It is a lesson that has ultimately shaped how I approach challenges and interact with others to this day, reminding me that sometimes, letting go of the small battles allows you to win the war for a better attitude and a more positive impact on the world. And maybe, just maybe, a little less "Goffrey" and a little more "Jeffrey" wouldn't hurt either.

The Day I Tried to Topple a Library Shelf

The library, for many, is a sanctuary: a hushed haven of knowledge, a tranquil escape from the cacophony of the world. For a restless, easily bored teenager like my past self, it was often little more than a vast, echoing vault of enforced quietude, populated by stern-faced librarians and the dusty scent of forgotten paper. And it was in this very setting, on a particularly dreary Tuesday afternoon, that I concocted one of my more spectacularly ill-conceived schemes.

It started subtly, as most bad ideas do. Homework was "done" (read: abandoned), and my mind was drifting. My gaze fell upon a particularly imposing set of shelves, groaning under the weight of countless biographies. They looked so heavy, so solid. And then, the thought, insidious and compelling, took root: Could I move it?

I wasn't trying to cause chaos, not really. There was no grand plan to send an avalanche of literature crashing to the polished linoleum. It was more a scientific inquiry, albeit one driven by pure, unadulterated boredom and a nascent, misguided sense of my own physical prowess. Could one lone, slightly spindly teenager, with a surge of adrenaline and a complete lack of spatial awareness, truly shift a structure designed to hold hundreds of pounds of printed matter?

I waited for the coast to be clear. Mrs. Gable (not her real name), the librarian, a woman whose mere presence commanded silence, was behind her counter, spectacles perched on her nose, lost in a book of her own. Perfect. I sidled up to the chosen shelf. Then, a quick glance around. Clear.

I braced myself, planting my feet, my hands perched on the cool, metal frame. I gritted my teeth, inhaled deeply, and pushed.

It was an embarrassingly pathetic effort. The shelf, a stoic monolith of knowledge, didn't even wobble. Not a millimeter. Not a whisper of a creak. It stood there, utterly defiant, mocking my puny strength with its immovable gravitas. I pushed harder, grunting slightly, my face probably turning an alarming shade of red. Still nothing. It was like trying to shift a mountain with a feather.

Then, the quiet shifted. Not a sound, just a sudden, profound awareness of being watched. I slowly turned my head, still braced against the unyielding shelf, and met the unwavering gaze of Mrs. Gable. She hadn't moved from her desk, hadn't made a sound. But those eyes, magnified by her glasses, conveyed a silent sermon of disappointment, bewilderment, and absolute certainty.

My blood ran cold. The brief, idiotic burst of energy drained from me, leaving behind a hollow pit of dread. There was no need for words. The sight of me, red-faced and straining against a library shelf, was indictment enough.

The walk to the principal's office was one of the longest journeys of my life. Each step echoed the stupidity of my actions. The principal, a perpetually weary man named Mr. Henderson (also not his real name), listened to Mrs. Gable's concise, damning report with a slow nod. He didn't erupt in anger; his disappointment was far more potent.

"So," he said, fixing me with his own tired gaze, "you tried to push over a library shelf?"

All I could manage was a meek, miserable nod. There was no plausible defense, no cunning alibi. The shelves had stood firm, but my reputation had crumbled.

The consequence was swift and unavoidable: three days of after-school detention. Each afternoon was a purgatory of ticking clocks and forced contemplation in a sterile, silent room. It gave me ample time to reflect on the physics of leverage, the surprising stability of industrial-grade shelving, and the keen observational skills of librarians.

I never again looked at library shelves without a new-found respect. They weren't just storage; they were steadfast guardians, unyielding bulwarks against the tides of youthful idiocy. And that Tuesday afternoon, a seemingly minor act of boredom-fueled mischief, taught me a surprisingly profound lesson: some things are simply not meant to be moved.

But funny enough, that silly incident also nudged me toward a different kind of realization. Maybe not about literal library shelves, but about ideas, especially when people try to put you in a box. Because while a shelf might be unyielding, your potential definitely shouldn't be.

We've all likely encountered someone who tried to define us, to box us in based on their preconceived notions. Maybe it was a well-meaning relative, a critical friend, or, more surprisingly, an educator. These moments, though often painful, can become powerful catalysts for growth and self-discovery.

I remember a teacher once telling me, with a disconcerting certainty, that I would never amount to anything more than a "dumb farmer." This pronouncement felt like a personal attack. What right did this person have to limit my potential based solely on my background?

It wasn't the only time I faced such limiting assumptions. Later, I was barred from taking a math test designed to identify the school's top math students. The excuse was a shortage of score sheets, compounded by the dismissive assertion that college wasn't in my future anyway. And then there was the guidance counselor who steered me away from college prep classes, reinforcing the idea that I wasn't "college material."

Looking back, these experiences highlight a critical point: No one should be allowed to dictate your path. Just because someone grows up on a farm or doesn't initially express a desire for higher education doesn't mean they are incapable of achieving great things. We all possess a unique set of talents and a capacity for growth that should never be underestimated.

The truth is that planting seeds of doubt can be incredibly damaging. Instead of nurturing potential, these individuals attempted to stifle it, based on narrow perspectives and preconceived notions.

The beauty of the human spirit lies in its resilience and its ability to defy expectations. Instead of internalizing these negative pronouncements, we can choose to use them as fuel for growth. Let the doubts and criticisms become the driving force that propels you forward.

Maybe you'll surprise everyone, including yourself. You might discover hidden talents, uncover unexpected passions, and carve out a path that far surpasses the limitations others tried to impose on you.

And even if your ambitions don't align with conventional success, taking control of your own narrative is a victory in itself. Want to go to college, even if others discourage you? Go for it! Even the act of striving, of pursuing knowledge and challenging yourself will benefit you in the long run. Those "college prep" classes, discouraged as they might have been, can provide invaluable skills and prepare you for future opportunities, regardless of your ultimate destination.

After taking the advice of my guidance counselor, I decided to switch things up and focus on something other than the usual academic classes. I chose not to take the SAT because I was told that college wasn't in the cards for me. It just didn't seem worth it to spend all that time and money on a test that wouldn't really benefit me in the long run.

Instead, I joined the Future Farmers of America (FFA) program and was ultimately chosen as the school's FFA president. Who would have thought that joining the FFA would lead me to become the school FFA president and County vice president? But there I was! Being a part of FFA allowed me to learn about agriculture, leadership, and teamwork. It gave me a sense of purpose and belonging in my school community and being selected as president was both an honor and a responsibility. And yes, I proudly wore my FFA jacket.

I got to roll up my sleeves and dive headfirst into re-building engines. From tearing them down to putting them back together, I learned a great deal about how these machines work and how to troubleshoot any problems that might arise. The hands-on experience was priceless; getting greasy and digging into the nitty-gritty of these engines really gave me a sense of accomplishment. Plus, working alongside my classmates made it all the more fun, instilling in me a sense of teamwork and camaraderie. (I won't mention the time we took the Vo Ag teacher's pickup for a winter drive and got it stuck in a snow-filled ditch. Imagine explaining why we were late for the next class!)

The Weight of Silt and the Lightness of Being

Junior year of high school is often characterized by a suffocating, internal lack of imagination. The world shrinks to the boundaries of the classroom, the judgment of peers, and the agonizing pursuit of an elusive, measurable self-worth. For me, the struggle was particularly sharp. Wrestling with the awkwardness of adolescence, I often felt less than, defined by what others said about me. I was drowning in the minor silt of teenage insecurity.

Then came June of 1977, and the devastating flash flood that ripped through Johnstown, Pennsylvania.

A call went out through our local church: volunteers needed immediately for the massive cleanup effort. I signed up less out of conscious concern for others and more out of a restless need to escape my own self-doubt.

The drive to Johnstown felt like crossing a threshold into a different geological age. The initial news reports had been horrifying, but they failed to convey the sheer magnitude of the water's destruction. Upon arrival, the air itself felt heavy, thick with the smell of wet earth, and something metallic and faintly organic, the scent of lives upturned.

Whole sections of the city looked like a child's toy set that had been violently shaken and dropped. Houses stood askew, their foundations compromised. However, the most pervasive evidence of the disaster wasn't the debris above ground; it was the mud below.

The other volunteers and I were assigned to a small residential street, our task simple and relentless: clearing the basements. We were a ragtag group of suburban teenagers and seasoned church elders, donned in rubber boots and cheap work gloves. We were given shovels and told to start digging.

The mud was silt—a thick, heavy composite of soil, heating oil, splintered wood, and the remnants of household goods. It was dark, slick, and bore a weight far exceeding that of normal dirt. It filled the basements in some cases from floor to ceiling, cementing itself to every surface and burying every possession.

The work quickly settled into a grueling, rhythmic monotony. We established a bucket brigade in the basement, passing heavy-gauge metal buckets hand-to-hand up the collapsing stairs and out into the sunlight, where they were dumped onto growing mountains of muck slated for removal.

Each bucket was a small, agonizing triumph. It weighed maybe thirty or forty pounds, and by the tenth bucket, my arms were tired. By the fiftieth, a dull, bone-deep ache had set in, blurring the physical reality of the mud with the emotional exhaustion of seeing the destruction.

Down in those basements, stripped bare of natural light and ventilation, we were excavating miniature tombs of memory. We didn't find treasures; we saw the heartbreaking indicators of lives interrupted: a mud-caked child's bicycle, a workbench still holding the outlines of tools long ago swept away.

I remember one basement, in particular, belonging to a middle-aged woman named Mrs. Henderson (not her real name). She stood silently by the curb, watching. When we finally cleared the last of the silt from a corner nook, we uncovered a small, wooden cradle, now ruined. Mrs. Henderson didn't cry; she just let out a small, sharp gasp that felt louder than any scream.

Those moments, witnessing such raw, material loss, began to chip away at the fortress of my adolescent preoccupations. What did my anxieties about my family's old car or my ill-fitting clothes matter when this woman was mourning the loss of a tangible connection to her past?

The only respite from the heavy work came three times a day, delivered like clockwork by the angels of communal care: the Red Cross and the Salvation Army. They came in unassuming vans and set up makeshift food lines on the muddy sidewalks. Their presence was a vital, comforting lifeline. Hot coffee, lukewarm Kool-Aid, and simple bologna sandwiches—these were sustenance, but more importantly, they were proof that the outside world had not forgotten Johnstown.

I remember the profound kindness in the faces of those volunteers—the unwavering smile of a Red Cross woman handing me a sticky donut, or the quiet, steady gaze of a Salvation Army man who simply nodded and asked if I needed more water. In my high school years, I struggled to find a sense of purpose. Here, covered head-to-toe in the stench of decay, I was seen, fed, and acknowledged as part of a crucial collective effort. The simple act of being routinely cared for—no strings attached, no judgments made—was profoundly healing.

In those moments, something inside me began to shift. The experience shattered my old assumptions about what it means to give and receive. The most transformative change, however, occurred inside my own mind. Before Johnstown, my definition of "poor" was absolute: lacking. My family didn't have some of the comforts others did, and this gap felt like something that defined my character.

But standing in that flooded city, helping people who had lost *everything*, the definition shifted.

We were poor, yes, but we had structure. We had walls that stood upright. We had dry clothes, plenty of food, and bedrooms waiting for us. Johnstown showed me the difference between *scarcity* and *obliteration*.

I realized, with a jarring clarity that shook me to my core, my family's financial struggles were not total destruction. For the people of Johnstown, the flood had wiped the slate clean—not just of wealth, but of history, security, and the simple, daily comfort of familiarity.

For a moment, all the petty insecurities that had plagued my junior year—the fear of rejection, the gnawing feeling of being less deserving—receded into insignificance. They were replaced by a startling appreciation of having a roof that didn't leak and a basement for the coal furnace, canned food, winter jackets, and yes, even a ping pong table.

I left Johnstown days later, my hands blistered, my back aching, and my clothes permanently stained with the rust-brown silt. I carried away no grand pronouncements about serving humanity, only a deep, abiding sense of empathy molded by the sight of sorrow and resilience.

The experience did not instantly solve the problem of my self-worth; those battles are long and complex. But it gave me a critical new metric. It taught me that worth wasn't measured by what I lacked, but by my capacity to labor alongside others, to share in sorrow, and to recognize the sacredness of what remained. The weight of those buckets of mud was immense, but the perspective they delivered was light, lifting me out of my own shadow and teaching me, for the first time, to appreciate the simple, profound gift of wholeness. I was poor, but I was intact, and in Johnstown, 1977, that felt like the greatest fortune of all.

> So, *remember this: Embrace the unexpected path. Sometimes, the detours we take, fueled by a refusal to accept limitations, lead us to discover passions and talents we never knew we possessed. Don't let anyone write your story for you. Pick up the pen and write it yourself. You might just surprise yourself with the incredible chapters you create.*

Chapter Five

From Unlikely Leader to Class President

The announcement hung in the air, but I barely registered it. Senior Class President? Me? It felt like a dream, a well-orchestrated prank. I was certainly no academic superstar, nor seasoned politician expertly playing the high school game. The idea that my classmates had entrusted me with leading them through our final year was, frankly, astonishing.

The surprises didn't stop there. Along with the unexpected presidential victory, I was also voted "Most Talented" and, even more bewilderingly, "Most Likely to Succeed." One particularly ambitious classmate even approached me, half-jokingly, and requested I relinquish the latter title to boost his college application odds!

Suddenly, the carefree days of junior year felt distant. The weight of the year ahead settled in. Choosing a prom theme, navigating the school bureaucracy to secure our senior class trip, and deciding how we wanted to leave our legacy—these were all weighty decisions that now rested on my shoulders, alongside the dedicated members of the student government. Looking back, I realize my classmates must have recognized leadership potential in me that I hadn't even acknowledged in myself.

Perhaps fueled by this newfound sense of responsibility and a healthy dose of audacity, a wild idea sprang to mind. "Why not invite President Jimmy Carter to our graduation?" I thought. After all, from one President to another... I penned a short, heartfelt letter, saying from one president to another, I am extending a warm invitation to you and Rosalynn to attend my high school graduation. Much to my astonishment, weeks later, a postcard arrived bearing the familiar insignia of the White House. While a physical visit wasn't possible, the warm message of encouragement nestled within was a powerful reminder of the possibilities that lay ahead. A former President (or his staff) taking the time to acknowledge our graduation—it was truly inspiring.

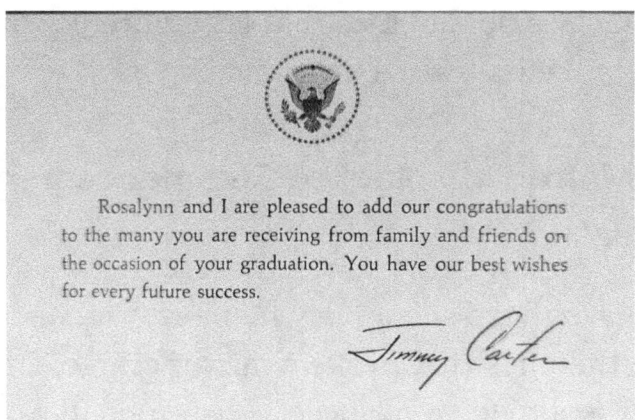

Rosalynn and I are pleased to add our congratulations to the many you are receiving from family and friends on the occasion of your graduation. You have our best wishes for every future success.

Jimmy Carter

Postcard from President Carter

The logistical challenges and the pleasant surprises, the culmination of my role as Senior Class President, was undoubtedly the opportunity to deliver the senior speech. The responsibility of representing my class, of offering meaningful words as we stood on the precipice of adulthood, was daunting. I dedicated weeks to crafting the perfect message, ensuring every syllable resonated with truth and offered genuine encouragement as we all prepared to embark on our separate paths.

I settled on a tree analogy, envisioning our foundational years as the roots, nurturing and supportive, and our future endeavors as the reaching branches, extending toward countless opportunities. When the day finally arrived, a potent mix of nerves and excitement bubbled within me as I stood before my peers. But as I began to speak, drawing from the wellspring of my heart, the anxieties dissipated. The tree analogy resonated, reminding us all of the strong foundation we had built together and the boundless possibilities that lay before us, reaching out like branches toward the sunlit sky.

From Mantle Mischief to Commencement Speaker

We all have those school day memories etched in our minds—the ones that evoke a mix of cringe, laughter, and maybe a touch of disbelief. I have one in particular that features a ceremonial mantle passing, a mischievous impulse, and an ex-girlfriend. It is a story that perfectly encapsulates the transformative power of time and how even slightly embarrassing past experiences can pave the way for the most unexpected of opportunities.

The ceremonial passing of the mantle is a tradition steeped in symbolism, a formal transfer of leadership and responsibility from one student to the next. As the outgoing student body president, my duty was to present this weighty symbol of authority to the incoming junior class president. And, being a teenager with a mischievous streak, I nearly derailed the whole thing.

The mantle, a heavy piece of fabric, was fastened with a pin strategically placed on the recipient's shoulder. The junior class president, as fate (and teenage angst) would have it, happened to be an ex-girlfriend. A devilish idea sparked in my brain: what if I just... jiggled the pin a little too enthusiastically? Just to see what might happen. It was undeniably childish, a momentary lapse in judgment fueled by a yearning to stir things up. Thankfully, my attempt at subtle sabotage was a spectacular failure. Nothing happened, no one noticed, and the ceremony continued without incident. But the memory lingers, a cringeworthy yet humorous reminder of my youthful immaturity.

And let's be clear, my academic performance in high school wasn't exactly stellar. I graduated comfortably nestled within the lower half —or perhaps even the lower third—of my class. I wasn't exactly setting the academic world ablaze. But graduating, regardless of class ranking, is an accomplishment. It signified the closing of one chapter and the exciting, albeit daunting, prospect of embarking on something new.

This "something new" demanded hard work and, crucially, personal growth, setting a new standard for myself that had little to do with grades and everything to do with personal improvement.

So, years after that ill-fated mantle ceremony and underwhelming graduation, imagine my utter surprise when I received a call from my old high school. They were inviting me to be the commencement speaker at the graduation ceremony. Me? The kid who graduated in the lower half of his class? The one who nearly committed a faux pas of epic proportions during a ceremonial mantle passing? They wanted me to address the graduating class?

The irony wasn't lost on me. It felt like the ultimate opportunity to silence any lingering doubts, to prove wrong to those who might have whispered that I wouldn't amount to much. It was a chance to share my journey, to inspire a new generation, and to demonstrate that success isn't always about starting at the top, but about the effort and dedication you invest along the way.

Of course, I said yes immediately.

The experience of returning to my high school, not as a mischievous student harboring questionable intentions, but as a respected guest, was both humbling and deeply rewarding. I had the honor not only of providing the commencement address but also of presenting the school with a plaque signed by me, which featured the American and Pennsylvania state flags that had flown on the Space Shuttle traveling over 1.7 million miles. The plaque was hung at the entrance to the school gymnasium for all to see and remains fixed on the school wall today.

It was a powerful reminder of the unpredictable nature of life and how even seemingly insignificant moments, like almost sticking a pin in an ex-girlfriend's shoulder during a mantle passing, can contribute to the intricate tapestry of our lives. It also proves that even those who start in the lower third can, with hard work, perseverance, and perhaps a touch of good fortune, find themselves standing on a stage, sharing their story, and inspiring others to chase their own unexpected dreams.

Chapter Six

The Beginning – My First Job

When I graduated from high school, I never imagined that my first job would be in a meatpacking plant. It wasn't the glamorous corporate internship or the chance to travel the world that most of my peers dreamed of. Instead, I found myself in a world of clanking machinery, pungent scents, and constant hum, most stressfully of all, those massive ammonia-based cooling systems. You had to be sharp, alert, and always on your guard. Ammonia leaks were no joke—one wrong whiff and you could be in serious trouble. It certainly wasn't a job for the faint of heart. Ironically, on scorching summer days, the chilling blast from those colossal coolers held a certain dark appeal. A quick, albeit slightly risky, step inside for a lungful of icy air offered a temporary and welcome respite from the sweltering heat.

My role was maintenance, which meant that I was responsible for troubleshooting everything from jammed conveyor belts to malfunctioning meat grinders to fixing vehicles. It demanded quick thinking and practical skills, such as welding in challenging conditions. The environment was messy, smelly, and physically demanding, but it unexpectedly cultivated a deep resilience and problem-solving prowess within me.

Despite the less-than-ideal circumstances, the role provided invaluable experience and a surprising sense of purpose. The hands-on experience, particularly the welding skills honed under pressure, proved invaluable, and eventually propelled me toward a more aligned sheet metal and welding job. It demonstrated how an unglamorous start can forge invaluable experience and a surprising sense of purpose.

Interestingly, this seemingly unconventional career path felt strangely familiar. Growing up on a farm, I was constantly surrounded by machinery, learning to diagnose problems and fix things on the fly. From repairing tractors to mending fences, I developed a knack for understanding how things work and the importance of keeping them in good repair.

In many ways, working at the meatpacking plant was just an extension of that upbringing. The scale was different, the machinery more complex, but the core principles remained the same. It is about understanding the mechanics, identifying the problem, and implementing a solution. The farm taught me the value of hard work and the satisfaction that comes from fixing things; those lessons translated directly to my current role.

This job might not have been what I envisioned after graduation, but it provided me with invaluable skills, a strong work ethic, and a deeper appreciation for the often-overlooked industries that keep our society functioning. While others might be chasing dreams in far-flung locations, I was finding my own sense of fulfillment right here, fixing machines and playing my part in the process.

And for now, that was a purpose I could be proud of. It may not have been glamorous, but it was real, it was necessary, and it was mine. And who knows what other surprising opportunities this path might lead to? The future, like the meatpacking plant itself, was full of unexpected possibilities.

Life has a funny way of throwing you curveballs, doesn't it? One minute you're wrestling with greasy machinery in the controlled chaos of a meat packing plant, and the next you're harmonizing in a choir, touring the heartland of the Midwest. Yep, that's precisely the unexpected journey I found myself on.

My job at the meatpacking plant started as the on-site maintenance guy; however, the day-in, day-out grind eventually left me yearning for something more. That's when I decided to take a break from the meatpacking job and enroll at the Rosedale Bible Institute for two semesters. The biggest draw? Without a doubt, the choir! The opportunity to tour the Midwest with a group of talented vocalists was simply too good to resist.

The tour itself was a unique experience. We stayed in different homes, hosted by members of the churches where we performed. As an introvert by nature, the constant stream of new faces and unfamiliar surroundings presented a bit of a challenge. I realized I had a choice: I could fade into the background, observing from a distance, or I could actively seek connection with these welcoming strangers.

You know how it goes. When you arrive at a new host home, the conversations often follow a predictable script: "Who are your parents? Where did you grow up?" The standard introductory small talk. But I decided to deviate from the norm and inject a little humor into the situation.

Instead of launching into my life story, I started asking, "Do you happen to have any cherry pie and ice cream?" You wouldn't believe it, but it actually worked! More often than not, the host families either had it on hand or were delighted to offer it as a treat. Maybe it was the novelty of the request, or maybe it was the undeniable charm of Midwestern hospitality, but that simple question became my unexpected icebreaker, transforming potentially awkward moments into fun, memorable experiences filled with laughter and delicious desserts.

Looking back, that unexpected detour—from the mechanical chaos of the meatpacking plant to the harmonious melodies of the choir tour—taught me invaluable lessons. It underscored the importance of embracing new expe-

riences, even when they push you outside your comfort zone. And it proved that sometimes, all it takes is a craving for cherry pie and ice cream to break down barriers and forge connections with people in the most unexpected ways. After all, you truly never know where life will take you next!

Married Life

"I do" at twenty years old? To some, it might sound like a whirlwind romance, a youthful indiscretion. But for me and Lauretta, marrying on my 20th birthday just felt... right. So right, in fact, that I ensured the date was permanently seared into my memory by strategically choosing my birthday as our wedding day. I figured if I ever almost forgot our anniversary, the deluge of birthday cards would serve as a helpful reminder. And let me tell you, that strategy has saved me from some serious marital faux pas over the years!

It may sound crazy to get married at such a young age, but some things you just know. We were young, carefree, and ready to take on the world together as husband and wife. And what a ride it has been! We've had our ups and downs, like any couple navigating the complexities of life, but through it all, I couldn't imagine doing it with anyone else but Lauretta. July 2025 is our 45th wedding anniversary. Time flies when you're having fun and enduring the occasional engine trouble.

Speaking of engine trouble, my curiosity, a trait I've had since childhood, didn't just disappear after high school. It just found new avenues to explore. After our marriage, we bought a brand-new Ford Escort, one of the first models they produced in the 1980s. The dealership boasted about its improved fuel efficiency, attributing it to the unique shape of the pistons. But they wouldn't divulge the specifics. That lit a fire under me. I had to know.

So, when the car reached around 40,000 miles, I decided to take matters into my own hands, literally. I removed the engine head to inspect those elusive pistons. It turns out that the difference was quite subtle—a simple design tweak. It would have been much easier if the garage had just told me what the change was, but the joy was in the discovery.

My curiosity didn't stop there. While I was knee-deep in engine grease, I started exploring other ways to boost fuel efficiency. That's when I came across an advertisement for a water injector. The idea was simple, in theory: spray water directly into the carburetor, based on engine speed, to create steam and increase compression, ultimately improving fuel economy. I had to try it.

I bought the kit, installed it, and even modified it so I could track the water injection rate. What could possibly go wrong? Well, plenty.

We lived in Maryland, and winterizing your car meant adding dry line antifreeze to remove any water from the gas tank. So, here I was, carefully removing water... and then actively spraying it directly into the carburetor.

One winter day, driving back from LaVale, about 20 miles from home, the engine started sputtering and struggling to climb a small hill on the freeway. The realization hit me like a cold wave—the water was freezing in the carburetor! I quickly shut off the injector, let the engine's heat melt the ice, and we were back on track.

But I wasn't ready to give up. I decided to add alcohol to the water injector tank to prevent further icing. Not knowing the proper ratio, I initially had periodic freezing problems. So, naturally, I went for the extreme solution: pure alcohol. What was I hoping to achieve? Ultimately, better fuel efficiency and lower operating costs. The problem was that the cost of alcohol likely outweighed any potential fuel savings. But, as they say, you never know until you try.

The good news? Despite my experimental engine modifications, the Escort managed to survive. And I learned a lot about engines, fuel efficiency, and the importance of a skilled mechanic. But more importantly, this whole escapade highlighted a key component of my marriage to Lauretta: a willingness to learn, and to laugh when things don't quite go as planned.

Because, just like adjusting the water injection rate on a Ford Escort, life is all about finding the right balance.

And I'm incredibly lucky to be doing it with Lauretta by my side, even if she sometimes rolls her eyes at my "innovations." After all, a little engine trouble is nothing compared to the joy of a lifetime spent together.

Welcomed into the World

A year after our wedding vows still echoed in our hearts, our lives were blessed with the arrival of our son, Eugene. From the moment his bright, inquisitive eyes focused on the world, we knew he was special. His sweet demeanor instantly stole our hearts, filling our home with a joy we hadn't known existed. He was, and continues to be, a gift beyond measure.

Eugene's middle name carries a profound significance: it pays tribute to Lauretta's late brother, Morris, whose life was tragically cut short in a car accident. This simple gesture ensures that his memory lives on within our family, a constant, poignant reminder of the love and loss that shape our lives and bind us together.

The circumstances surrounding Eugene's birth were, to say the least, unconventional, a beautiful reflection of Lauretta's unique upbringing. Before our marriage, Lauretta worked alongside her father in his medical office, a haven for a rural Amish community nestled in the heart of Ohio. The Amish community, traditionally hesitant about hospital births, preferred a more intimate and familiar setting. To accommodate their needs, Lauretta's father and another doctor established a special birthing house, meticulously equipped with essential medical tools, all within a space notably devoid of electricity.

Having witnessed countless childbirths alongside her father in this setting, Lauretta experienced firsthand the compassionate, personalized care provided. She felt strongly that she wanted Eugene to be brought into the world in the same way. I, however, was initially apprehensive, being accustomed to the modern conveniences and seemingly readily available support of a hospital. However, after carefully researching the success rates of births at the birthing house versus the local hospital, my concerns were eased. I readily agreed with Lauretta's heartfelt wish, understanding the profound connection she felt to this intimate setting.

Eugene made his grand entrance around 7:15 pm. I will never forget the surreal experience of the delivery. The doctor, illuminated by what looked like a miner's light, carefully guided Eugene into our arms. It was a scene straight out of another time; a moment suspended between tradition and modernity. Adding to the uniqueness of the experience, our Amish host lovingly prepared a mouthwatering four-course meal for us, a gesture of warmth and welcome that remains indelibly etched in my memory. It is funny how certain details stay with you; the flickering light, the gentle hands, and the aroma of freshly baked bread, all woven together into a tapestry of love and community.

As we watched Eugene grow, we were constantly reminded of the beauty and wonder of life. His infectious smile and charming personality brightened every day. It was an utter delight to witness his first tentative steps and hear his first precious words. We cheered as he scored his first soccer goal, won his first high school tennis match, wrestled with determination, and played the French horn—just as I had in my own high school days. We beamed with pride as he walked across the stage at both his high school and college graduations. We felt the same joy watching him marry his beautiful bride and, later, seeing his son inherit so many of the traits we had cherished in Eugene himself. We feel incredibly fortunate to have Eugene in our lives, filling our family with love and laughter with each passing day. He is a constant reminder of the profound beauty and wonder of life, a gift we cherish beyond words, a legacy of love woven from tradition, remembrance, and hope.

Living just a short walk from my grandfather's house made visits a cherished part of our lives. I still vividly recall the days I took Eugene, then just a year old, to his great-grandparents' place, pulling him along happily in his little red wagon. Those afternoons were often spent in my grandfather's bustling wood shop, where we'd lend a hand, whether it was crafting sturdy flower planters or whatever other project he had on his mind that day. The scent of sawdust and fresh-cut wood always filled the air, a comforting aroma of shared work and quiet compan-ionship. And the best part? After we'd packed away the tools and swept up the shavings, my grandfather would inevitably invite us into the house for a well-deserved bowl of ice cream, a delicious end to a perfect day.

My Wild Ride Out of the Meatpacking Plant

Life's funny, isn't it? You think you're stuck on one path, destined to repeat the same routine, and then BAM! Something shifts, and you find yourself hurtling down a completely different road. My journey out of the meatpacking plant was exactly that kind of wild ride. I was a fixture there, bundled up against the chill, working hard and knowing what to expect. It was a steady job, a familiar rhythm.

But let's be honest, the daily grind of the meatpacking plant takes its toll on you. The cold seeps into your bones, the work is physically demanding, and the risk of injury is always lurking. So, after what felt like a lifetime of wrestling with frozen beef, I decided to embark on a new adventure as a salesman for voltage surge suppressors. Yep, you read that right. From meat to voltage sparks, from frozen to fired up. It might sound like a crazy jump, and honestly, it probably was. Leaving the security (and let's face it, the predictability) of the meatpacking plant was one of the scariest decisions I've ever made.

Don't get me wrong, trading a meat hook for a multimeter brought a whole new set of challenges. The physical risks may have decreased significantly, but the stress of running a business is a completely different beast. There were days (and nights!) filled with doubt, wondering if I had made the right call.

I learned the ins and outs of the industry, honed my sales skills, and discovered a passion for protecting people's valuable electronics.

When Pride Takes a Backseat

Sometimes life throws a curveball, transforming what should be a period of blissful new parenthood into a whirlwind of stress and uncertainty. That's exactly what happened to my wife and me. Shortly after our son was born, my job situation took a steep dive, leaving us scrambling to make ends meet. The chaos of new parenthood was amplified by the looming shadow of financial hardship.

We entered survival mode. Luxuries were the first casualties, replaced by a laser focus on budgeting and finding savings wherever possible. It was a nerve-wracking time, a constant battle against the rising tide of bills and unexpected expenses. Rent and car payments loomed large, and we were always playing catch-up, struggling to keep our heads above water. Tough conversations became commonplace as we dissected our spending, prioritizing which bills to pay and where to cut back further.

One particular night remains etched in my memory. Dinner consisted of boxed macaroni and cheese—the kind you simply boil and stir, a far cry from the gourmet meals we once enjoyed. Washed down with water, it was the last food in the house, save for baby food. The weight of our situation was palpable, a heavy burden borne on empty stomachs.

Later that night, a knock on the door broke the silence. Standing there was someone I wasn't particularly fond of, accompanied by his young son. He awkwardly explained that they had extra food and wanted to share it.

How did he know? We had tried so hard to keep our situation private. Was it divine intervention? Perhaps. But I believe there was a more profound lesson at play.

Suddenly, I was faced with a critical choice: allow my pride to dictate my actions or swallow it and accept his offer. The internal battle was fierce. Humiliation threatened to overwhelm me. The thought of accepting help, especially from him, felt like a personal defeat.

But then I looked back at the faces of my wife and son, their needs outweighing my ego. I knew what I had to do.

With a deep breath, I thanked him, my voice thick with emotion. I swallowed my pride and accepted the food. It wasn't just a meal; it was a lifeline, a symbol of hope, and a profound lesson in humility. It taught me that in times of desperate need, pride can be a dangerous luxury, far more costly than a humble acknowledgement of vulnerability.

That night, I learned that sometimes help comes from the most unexpected places, reminding us that even in the darkest of times, kindness and compassion can still shine through. Focusing on the good of people rather than the negative became a new mantra for me. The macaroni may have been gone, but the lesson learned that night will stay with me forever.

An Unexpected Journey of Learning and Self-Discovery

Life rarely sticks to the script we initially write for it. Honestly, sometimes the most profound and genuinely fulfilling adventures aren't found on those well-trodden paths we think we should take. Instead, they pop up from unexpected detours, when we're brave enough to just step into the unknown. Shortly after the macaroni experience, my own path took precisely such a sharp turn when I found myself saying "yes" to a role as a sheet metalist and welder for a company that manufactured high-voltage test equipment.

It was a pretty dramatic departure from anything I had done before. The intense heat and the precise nature of this kind of welding were a stark contrast to my previous, rather basic welding experience at a meatpacking plant. But there was this sheer allure of mastering new skills and really pushing the boundaries of what I thought I could do, and it was just utterly irresistible. This wasn't just any welding, by the way; I was fabricating components and cabinets for these formidable pieces of high-voltage test equipment. These giant machines are critical tools that power companies use to meticulously test the integrity of transformers on their vast power line networks. We're talking towering pieces of machinery, custom-mounted on robust flatbed trailers, designed to produce an astonishing one million volts! Indeed, every single phase of testing these powerful systems demanded an unwavering level of caution and meticulous attention to safety.

The work was definitely demanding, testing both my physical strength and mental focus. However, despite that, I found myself becoming more and more captivated by the hands-on nature of the craft. Mastering the art of fusing intricate metal pieces together, creating the very structures that formed the backbone of high-voltage test equipment, was incredibly rewarding. Seeing my hard work literally come to life, knowing it played a crucial role in ensuring the smooth operation of critical systems, filled me with a tangible sense of purpose. This experience not only expanded my skill set but it also instilled a deep appreciation for the importance of precision and dedication.

Just as I was starting to feel comfortable in my new role, life threw another surprise my way. This time, it was in the form of a chance conversation during a break. I struck up a chat with Lloyd, the group lead and coworker. As we delved into deeper topics, he cautiously opened up about his experiences as a radio technician in Vietnam. He rarely spoke of it, but when he did, the sheer weight of his experiences was palpable. He recounted the intensity and chaos, emphasizing the importance of staying calm and focused in order to maintain communication lines under immense pressure. He downplayed his role significantly, but it was clear to me that he had made an invaluable impact.

Noticing my burgeoning interest in electronics—something I hadn't really given much thought to before—Lloyd unexpectedly suggested I consider taking some evening courses at the local community college to develop my skills. This suggestion planted a real seed, stirring something within me. I confessed to Lloyd my high school experiences, admitting the self-doubt that had been instilled by teachers who had subtly painted me as someone not really suited for higher education—someone, in their eyes, destined to remain a farmer.

Then came another surprise, one that would truly alter my perspective forever. "Let's enroll in college together," Lloyd declared. I hesitated, reiterating my ingrained concerns about academic inadequacy. He simply insisted he would take the class *with* me, offering that quiet, unwavering support I didn't even realize I craved. And so, just like that, we both enrolled in Digital Electronics 101.

There we were, side-by-side, wrestling with the intricacies of circuitry and the logic of programming. The experience was surprisingly enjoyable, especially having someone to collaborate with and bounce ideas off. We learned together, struggled together, and ultimately, aced the course together.

This success ignited a fire within me. It truly struck me that perhaps I wasn't the person those high school teachers had envisioned. It was a real wake-up call, a powerful realization that I needed to take control and be more proactive in shaping my personal and professional life. It became incredibly clear that Lloyd, a man likely capable of teaching the course himself, had taken it with me not for his own benefit, but to provide the vital support and encouragement I needed to break free from my self-imposed limitations.

My journey from being a welder to a part-time electronics student is a powerful testament to the influence of unexpected opportunities and the profound impact of mentorship. One chance conversation, one act of simple kindness, can genuinely change the trajectory of a life, proving that sometimes, the most rewarding paths are indeed the ones we least expect. It is a potent reminder that learning is a lifelong journey, and with the right support, we are all capable of far more than we realize.

This experience taught me a valuable lesson that has guided me throughout my career: to always be on the lookout for others who need that extra boost or a figurative hand to hold, to provide encouragement and mentorship, and to help them unlock their own hidden potential. Because sometimes, all it takes is a spark of belief to ignite a circuit of possibilities.

From Fourth Grade to Magna Cum Laude

Life isn't always a straight line. For me, the path to a fulfilling engineering career and a Magna Cum Laude degree was paved with unexpected detours, unwavering support, and the sheer grit to overcome a past riddled with setbacks. My journey is a powerful testament to the transformative potential of education and the enduring power of second chances.

My academic journey truly began with a spark that ignited when I started attending community college. After years of feeling like I just wasn't good enough academically, I finally proved to myself that I could succeed. This realization, coupled with a burning desire for career advancement, led my amazing wife, Lauretta, and me to a pivotal decision: I would leave my current job and go all-in on pursuing a four-year engineering degree. It was a tough call, for sure, but we both knew deep down it was the right move for my career and our future together. Lauretta's unwavering support really fueled my confidence, setting the stage for a challenging but super exciting new chapter.

However, the past definitely cast a long shadow. When I started applying to engineering colleges, I faced the harsh consequences of a pretty lackluster high school record. I honestly thought my passion for tech would be enough to just get me in, but boy, was I wrong. Poor grades and the absence of an SAT score presented some significant hurdles. The dream of becoming an engineer suddenly seemed to be slipping away. But instead of just succumbing to discouragement (which would have been easy to do!), I decided to embark on a mission to strengthen my application, determined to show my unwavering commitment to the field.

My perseverance paid off big time when I received an acceptance from Capital College, a prestigious private institution in Laurel, Maryland, into its Telecommunication Engineering program. There was a bit of a catch, though: I had to prove myself during my first year. With no scholarship eligibility, I ended up taking out student loans and snagging a part-time job to make ends meet.

I applied for a position in the college's maintenance department, thinking my farm background and mechanical aptitude made me a perfect fit. Rejection, however, led me to an unexpected opportunity: a part-time job in the school library. (No, I didn't try to move the bookshelves.) This seemingly less desirable position turned out to be a total blessing. I was so grateful to be in a cozy environment during extreme weather rather than out in the elements—trust me, that was a huge win!

Balancing the responsibilities of being a husband, a father, a full-time student, and a part-time employee was undeniably demanding. The library job provided such a supportive environment and the chance to help fellow students find the resources they needed. I honed some very valuable time management skills and learned to prioritize effectively.

Freshman year brought a new wave of anxieties. Hearing classmates casually discuss their advanced preparatory classes, including calculus, triggered a full-blown panic! I enrolled in Calculus I, diligently took notes, and dedicated myself to studying like crazy. Despite lacking the high school preparation of my peers, I not only passed but earned an "A." This triumph completely defied a long-held belief in my mathematical abilities and underscored the pure power of determination.

The financial strain was significant, too. Paying for childcare, with no family nearby to help, added a lot to the pressure. It wasn't easy seeing a chunk of our paycheck disappear every month but knowing that our little guy was in good hands made it all worthwhile. Despite the financial sacrifices, we always prioritized our son's well-being and development.

Even as a full-time student and holding a job, I didn't forget to prioritize time with my son. As he got older, our shared adventures at "ham fests" (computer shows, not meat, thankfully!) and our weekly bike rides to Dunkin' Donuts fostered a strong bond. Discovering Eugene's passion for card collecting, we embarked on a shared hobby, attending card shows and autograph events, accumulating over 200,000 cards and countless autographs.

In my junior year, a pivotal decision completely reshaped my academic trajectory. I decided to switch majors from telecommunications to electrical engineering. This shift was driven by a deeper passion for electronics, the hands-on nature of the field, and the vast possibilities for innovation. I just recognized that electrical engineering aligned so much better with my long-term career aspirations.

My telecommunications coursework did prove to be invaluable when I secured a Co-Op position at ARINC in Annapolis, MD. There, I gained practical hands-on experience testing critical ground-to-air communications for commercial aircraft, which allowed me to develop crucial skills in programming, troubleshooting, and rigorous testing of complex communication systems. A significant part of the work involved tracking these commercial aircraft across the country, constantly monitoring the real-time performance and reliability of their vital communication links, ensuring seamless and safe operations. This immersive experience truly bridged the gap between theoretical knowledge and practical application, solidifying my expertise in critical aerospace communication technologies. Seriously, it was like a masterclass!

My success at ARINC then led to another Co-Op opportunity, this time at Litton Systems in College Park, MD. At Litton, I immersed myself in military systems, space projects, and commercial ventures, further honing my skills and expanding my knowledge base. This experience underscored the incredible importance of communication and strategic planning in demanding environments.

My performance was so impressive that I was offered a full-time position at Litton Systems while still completing my senior year! The opportunity to contribute to space projects was a particular highlight, offering firsthand exposure to cutting-edge technologies.

The culmination of years of hard work, sacrifices, and unwavering dedication arrived on graduation day. I walked across the stage, a four-year degree in hand, graduating Magna Cum Laude. A profound sense of accomplishment washed over me. I had not only defied those who doubted my abilities, but I had also soared to achieve a significant milestone. It just goes to show that it is never too late to reinvent yourself and chase those dreams, no matter how bumpy the road might seem.

> *My story is a powerful reminder that with passion, perseverance, and unwavering support, anything is possible. My journey from fourth-grade struggles to Magna Cum Laude is an inspiration to anyone facing adversity and a testament to the transformative power of education. And, as I reflect, I offer this piece of advice: Not applying myself in high school resulted in higher stress levels during college. Actions have consequences! This hard-earned lesson serves as a final reminder that the choices we make today shape the future we create tomorrow.*

Finding Strength Amidst Unimaginable Loss

Have you ever experienced something that completely shattered your perspective, forcing you to re-evaluate everything you thought you knew? These experiences shake us to the core, confront us with our deepest fears, and demand a resilience we didn't know we possessed. For me, that moment arrived in 1998 when tragedy struck my family, leaving an indelible mark on my life and forever changing the way I view the world.

That year, our family farm in Meyersdale, Pennsylvania, became the target of not one, but two tornadoes. Seriously, two! The sheer force of nature ripped through our lives, uprooting trees, demolishing buildings, and leaving a landscape of utter devastation in its wake. The chaotic aftermath felt totally surreal. While the community immediately rallied to help, the true weight of the disaster was yet to be realized.

Destroyed barn after two tornados in two days

Damaged house from the tornados

Amidst the chaos, the support from our family, friends, and neighbors was a literal beacon of hope. Organizations like the Red Cross and the Salvation Army stepped up, offering vital assistance—food, emotional support, clothing, household items, and financial aid. Their presence was a real source of comfort and security during our darkest hours.

But the devastation caused by the tornadoes went beyond the physical destruction of our home and livelihood. I was working at Litton Systems in College Park, Maryland, when I got the shocking news that the farm had been hit by the two tornadoes. Without a second thought, I immediately embarked on the four-hour drive back to Pennsylvania, knowing that family always comes first, no matter what.

As I began that long journey, my phone rang again, bringing news that would forever alter the course of my life. My sister's husband and daughter had tragically passed away from carbon monoxide poisoning. Can you believe it? Caused by a generator running in a confined space in the basement. Despite repeated public warnings—seriously, they're everywhere—my brother-in-law had disregarded the dangers, placing a generator in the basement just to keep the refrigerators running during the power outage. The drive suddenly felt endless, each mile a painful, gut-wrenching reminder of the world I was entering without two beloved family members.

Desperate for any kind of distraction, I somehow found myself reminiscing about something totally bizarre. I clung to a fragmented memory of a tornado hitting my uncle's farm when I was a boy. I vaguely remembered my parents telling me that a *tomato*—at least that's what I thought they said—destroyed my uncle's farm. It was such a nonsensical thought, but in a strange way, it actually helped me stay grounded amidst the overwhelming grief. The sheer absurdity of it even brought the movie "Fried Green Tomatoes" to mind. Looking back, it seems silly, almost laughable, but it served as a temporary, weird shield against the agonizing reality.

Upon arriving at my sister's house, I was the very first family member to appear on the scene. The house was still a whirlwind of emergency personnel and police activity. Advised not to enter, I just stood on the periphery, numb with disbelief. My sister, who worked night shifts at a nursing home, had returned home to a house filled with carbon monoxide and deceased loved ones. She was, understandably, rushed to the hospital. It was all so overwhelming that I couldn't even recall her last name when questioned by the police—just goes to show you the profound impact that extreme stress can have on your mind.

Finally arriving at the farm, I was met with a scene of utter devastation. Debris was scattered across the fields, and the house and barn were in shambles. The missing roof, the demolished barn, the familiar sandbox now buried beneath fallen trees—all remnants of a cherished childhood were just... gone. Even the beloved Honda 750, a symbol of my youth and freedom, was completely destroyed. It was a stark reminder that some things, once gone, are simply irreplaceable.

Looking back, it is hard to articulate the sheer weight of that period. It was not just the physical destruction or even the initial shock of the tornadoes. It was the gut-wrenching, unshakeable grief from losing family members in such a preventable, tragic way. It makes you realize how fragile life is and how quickly everything can change.

But here's the thing about unimaginable loss: it forces you to dig deep. It strips away the superficial and confronts you with what truly matters. I learned that strength isn't about not breaking down; It is about finding the tiny glimmers of light in the darkest places and somehow, putting one foot in front of the other. It is about leaning on your community, accepting help, and finding a weird kind of peace in the knowledge that you survived, even if you're forever changed.

Seeing those familiar emblems—the Red Cross and the Salvation Army—rolling up to offer support felt like a punch to the gut, but in the best way possible. Suddenly, my mind zapped back to 1977 to the Johnstown flood. I was younger then, one of the volunteers with a shovel in hand, helping clean up the mess. And boy, was I ever grateful for a hot coffee or a simple meal handed out by those very same folks.

This time, though, the roles were completely flipped. I wasn't the one covered in mud, thankful for a handout; I was the one experiencing that unsettling disruption, that sudden, overwhelming need. And when help appeared, it was this quiet, almost overwhelming relief that just washed over me. It is a very humbling pivot, let me tell you. Realizing that the very same sustenance I once witnessed being given to sustain the helpers was now being extended to sustain me? Yeah, that hits different. This full-circle moment has just cemented my appreciation for these tireless organizations. It is not just a mission for them; It is like an unbroken chain of human care, always there, ensuring that in our most vulnerable moments, no one—whether you're a victim or even a volunteer—is left without the vital nourishment and dignity they provide.

This experience reshaped my entire perspective. It taught me profound lessons about resilience, the preciousness of life, and the enduring power of human connection. While the pain of that event never fully disappears, it is transformed into a quiet understanding of what it means to truly live, to appreciate every moment, and to find strength even when you feel like you have none left. It was a brutal lesson, but one that ultimately showed me just how much capacity the human spirit has to endure.

And ironically, right when I was experiencing this unexpected support from strangers, I was also deep in the trenches of another kind of crisis, one that demanded a completely different kind of strength from me. My sister was reeling from an unimaginable loss, and I just knew I had to step up. Planning the funerals, navigating all that arduous process—despite my own overwhelming grief, my focus was squarely on her. I had to be there, a steady presence amidst all that darkness. It meant suppressing my own pain, putting on that brave face, pretending that everything was, well, alright. It was emotionally draining, no doubt about it, but seeing the comfort and support I provided to my sister? That made every single ounce of it worth it.

That whole experience taught me such a profound lesson about the human capacity for resilience. It truly forced me to learn how to compartmentalize my emotions and to prioritize the needs of others even when I was struggling myself. Trust me, it wasn't easy, and the emotional toll was significant. But I learned the importance of being a pillar of strength for loved ones when the chips are down, when life throws its absolute worst at you. While the scars from that year certainly remain, they serve as a constant reminder of the power of family, the incredible kindness of strangers, and the unwavering strength we find within ourselves when faced with unimaginable loss. It was a period when the sky truly felt like it had fallen but from the rubble, I learned the enduring power of love and resilience.

Chapter Seven

A Journey of Perseverance and Unexpected Turns

L ying in the grass, staring into the vastness of the universe, my dream began. As a child, I yearned to be a part of the organization exploring the stars, though I didn't even know it was NASA at the time. Getting into NASA, however, proved to be a goal in a league of its own. It took 12 years of unwavering hard work and perseverance before I finally joined the ranks at the Johnson Space Center in Houston, Texas. My journey was a testament to the power of seizing opportunities and embracing challenges, even when they led down unexpected paths.

My initial path wasn't a direct shot to Houston. I accepted a full-time position as a Reliability Assurance Engineer (RAE) at Litton Systems, a reputable company where I knew I could build crucial skills. Little did I know that technical problem-solving, management, time management, teamwork, and attention to detail honed at Litton were laying the very foundation I needed to one day reach for the stars.

As an RAE, my primary responsibility was to ensure the smooth operation of the systems assigned to me. This involved reviewing parts, scrutinizing drawings, addressing emergent issues, and identifying system-wide problems. I remember one particularly challenging problem that left me completely stumped. Experiment after experiment yielded no answers. Then, one morning, (apparently, my mind kept working overnight) the solution crystallized in my mind. It was a Eureka! moment. I immediately tested my theory, and sure enough, it was the answer. This experience reinforced the importance of perseverance and trusting my intuition.

Opportunities for growth continued to present themselves, and I was entrusted with the pivotal role of Qualification Test Director for a crucial Navy aircraft avionics upgrade project. This was a responsibility of immense scope, demanding relentless coordination not only with our internal engineering and technician teams, but also directly with prime contractor leadership. My ultimate goal was to orchestrate a comprehensive testing regime that guaranteed the absolute safety and uncompromising functionality of the new avionics system. The testing itself was a rigorous gauntlet, pushing the sophisticated avionics to their limits and beyond. We spent countless hours in specialized facilities where the environmental chambers—those imposing steel boxes—simulated the extreme conditions of flight, subjecting the systems to bone-chilling cold, scorching heat, and the vacuum of high altitude. At the same time, massive vibration tables relentlessly shook, rattled, and rolled the equipment, replicating everything from engine resonance to the violent shocks of carrier landings. These were often long nights, illuminated only by the glow of monitors displaying real-time data, as we meticulously observed every sensor reading, every circuit response. Troubleshooting became a way of life, dissecting anomalies, debugging software glitches, and recalibrating hardware until every parameter was within specification. The demands were immense, testing not just the system but also our endurance. Yet the profound feeling of accomplishment when every test suite passed and every integrated component performed flawlessly was truly exhilarating—a highlight that solidified my commitment to ensuring the highest standards of aerospace safety and performance.

Following the Navy project, I was appointed Qualification Test Director for a secure communication system. This role demanded careful attention to detail as I oversaw the testing of cutting-edge technology to ensure its readiness for deployment. The challenge was considerable, but the opportunity to work on such advanced technology and contribute to secure communications was immensely rewarding.

Then came the opportunity to step into a leadership role. My boss, recognizing my potential, offered me the position of Engineering Project Manager (EPM) for systems used on LA-class submarines and Navy aircraft. These systems were critical for early detection and warning during missions. This was an opportunity to truly showcase my leadership skills, and despite the initial nervousness, I embraced the challenge with enthusiasm. The unwavering support of my boss and the acknowledgment of my hard work fueled my determination.

These leadership experiences, coupled with the technical expertise I had gained, proved crucial preparation for my time at NASA. From leading group projects to making tough decisions under pressure, I was constantly pushed out of my comfort zone, fostering growth and development.

However, the path wasn't always linear. The Chief Operating Officer of Litton called me in to serve as a corporate troubleshooter, investigating unusual activities in various company Divisions. This was a testament to my reputation for fixing things. The role was high-pressure, often requiring quick thinking and decisive action. I was once sent on a one-way flight, promised a return ticket only after I cracked the case.

One particular troubleshooting effort kept being extended, leading to constant rescheduling with my wife. In a moment of pure frustration, I asked my boss to call her and explain the situation—and he actually did! While embarrassing, it highlighted the dedication required in these roles. In the end, she understood, and we were able to reschedule.

On another assignment, I was given just three days to identify the cause of failure for a critical item used by the military, a failure that was costing the division significant dollars per week and risking their performance rating. With the Litton COO and a high-ranking military official arriving to demand answers on the third day, the pressure was immense. Fortunately, I identified the cause quickly, presented my recommendations, and resolved the issues, getting the project back on track.

Ultimately, my journey at Litton brought me full circle to my childhood dream. I was assigned as Task Manager for the manufacturing and delivery of the Space-to-Space Communications System (SSCS) to NASA. This system, crucial for astronauts communicating with the Space Shuttle and the International Space Station during Extra-Vehicular Activities (EVAs), was a big deal. My role required frequent trips to the Johnson Space Center (JSC), where I interacted with my NASA counterparts. I was later presented with an SSCS plaque by JSC for my contributions, and a flag flown on Shuttle Atlantis STS-106, a moment of immense pride.

However, the SSCS system wasn't performing as expected, and pinpointing the root cause proved challenging. During one visit to JSC, I felt the issue lay in the JSC design, not our manufacturing process. Frustration reached a boiling point, and in a moment of exasperation, I suggested to the Division Chief that he should hire me to resolve these types of issues. It was a bold, audacious statement, but I knew my skills were needed.

SSCS *Plaque*

He simply suggested I submit my resume. I did, and two weeks later, I received a postcard from NASA, thanking me for my interest but stating that there were no opportunities available at that time. Despite feeling confident, my dream seemed to slip through my fingers.

Working for Litton was never solely about the daily grind; the company also emphasized professional development and product support through various conferences, often held in inspiring locations. One such memorable event took us to the picturesque town of Breckenridge, Colorado. This wasn't merely a business trip for me; my wife and son enthusiastically joined me on the journey, eager to explore the mountains alongside me. During some limited free time, we seized the opportunity for an exhilarating adventure, strapping ourselves onto snowmobiles. We carved a thrilling path through dense evergreen trees and over undulating hills, the crisp mountain air filling our lungs, until we reached the majestic expanse of the Continental Divide. It was an absolutely unforgettable experience, blending work and family adventure in the most spectacular way.

But the story wasn't over. My journey had just begun.

The lesson is that patience is a virtue when it comes to making decisions. Although I thought I had exhibited fortitude in preparing to work for NASA, I would have to stretch the limits of my patience if I truly wanted to achieve my dream.

Chapter Eight

The Power of Decisive Action

L ife's a funny thing, isn't it? It is jam-packed with these moments, these crossroads, where you've got to pick between what's comfy and what's totally unknown, between sticking with what's safe and really going for it. And often, these moments just appear out of nowhere, demanding that you make a quick choice, take a real leap of faith. I learned that firsthand when something I had dreamed about for ages suddenly landed right in my lap in the form of a phone call.

About six months after our initial conversation, my phone rang, and it was the Division Chief. He didn't waste any time. He told me NASA's Johnson Space Center had the green light for three "critical hires," and he was authorized to offer me one of those spots. The catch? This authorization wasn't going to last forever, and I needed to make up my mind that morning. He needed an answer, or he'd just move on to the next person in line. No pressure, right?

Now, this wasn't just another job offer. This was a potential game-changer. Saying "yes" meant packing up our lives in Maryland and moving all the way to Clear Lake, Texas. It meant our son, who was just starting his sophomore year at the University of Maryland, would be staying behind. It meant leaving behind our familiar crew of friends and trying to build new connections from scratch in a whole new state. Talk about a lot to chew on!

The weight of that decision felt huge. But deep down, like way, way down, I just knew that this was the chance I would been working toward my whole life. So, less than two hours after that call, I picked up the phone and accepted the position. I didn't even ask about the salary or benefits—I was just confident those details would sort themselves out. I saw how quickly this window of opportunity could close, and I just knew I had to grab it.

Here's the thing: the world isn't going to wait around for everything to be perfect. Opportunities hardly ever show up with a big bow on top, or with a clear, straight path laid out for you. More often than not, we just have to weigh up what we could gain against what we might risk and then trust our gut.

Now, this isn't about blindly leaping at every single chance that comes your way. It is about spotting those moments that truly click with what you're aiming for, the ones that align with all the hard work and passion you've put in. It is about knowing the difference between taking a smart, calculated risk and just making a reckless gamble.

Taking those well-thought-out risks and seizing opportunities can lead to amazing success. On the flip side, trying to force a move when it just doesn't feel right? That can lead to mistakes and, ironically, missed opportunities down the road. It is a delicate balance, this dance between being patient and knowing when to act decisively.

It is like that old saying goes, "Don't wait until all the stop lights are green before starting to cross the city." Life rarely gives us perfect conditions. We just have to learn to navigate the uncertain bits, trust our own judgment, move in faith, and keep moving forward with a bit of courage and conviction.

My whole journey to NASA really hammered home the importance of being ready, staying focused, and being prepared to seize that moment when it finally arrives. So, stay patient, keep your eyes on the prize, and when that door swings open, don't even hesitate. Just walk right through it. The rewards might just be out of this world!

As you'll find out, I applied these very same principles throughout my entire NASA career. And that just goes to show how incredibly important it is to take stock of what we learn along the way and use those lessons to navigate all of life's crazy twists and turns.

Chapter Nine

How Music Training Shaped a Journey

For me, music has always been way more than just a bunch of notes scribbled on a page. It is this incredibly powerful thing that connects people, lets us express ourselves, and surprisingly, even helps us step up and lead. Looking back, I can honestly say that my early music training laid the groundwork for a truly vibrant journey, filled with many opportunities to serve, inspire, and make some truly unforgettable memories. From leading worship services to getting super creative with quirky props for children's ministry, this whole adventure has been completely unexpected and wonderfully rewarding.

My first big dive into this world involved leading Sunday morning worship services. It meant carefully selecting songs that resonated with the pastor's message and then guiding the whole congregation in collective praise. And trust me, it wasn't always perfectly smooth sailing. There were those moments, like the occasional, shall we say, "disconnect" with the organist—but honestly, those challenges just pushed us to collaborate more and helped me get a much deeper understanding of how musical arrangements really work. The pressure to be prepared and totally in sync ultimately made the whole worship experience better.

However, my journey took an unexpected turn. I was asked to lead a children's performance of "Cool in the Furnace." Yep, children's ministry! This was totally uncharted territory for me, but I figured, why not? I jumped in with both feet, full of enthusiasm and a surprising burst of creativity. The absolute highlight of that whole experience? Building this ridiculously dynamic prop: a fake thermometer, complete with tubes, rolled-up red paper, and even a LEGO motor to make the temperature "rise." It sounds like such a small detail, but man, it totally ignited the audience! It just goes to show, sometimes the most impactful moments come from the simplest, most imaginative touches.

The move from Maryland to Texas was a bittersweet moment, made even more poignant by a farewell concert that felt like the perfect culmination of years of shared memories with church members. I sang solos that really marked my journey with that congregation, and it was my chance to express some profound gratitude, reflecting on the powerful bond that music had forged within that community. As the melodies filled the air, my mind drifted back to a very special time when my son and I rebuilt the church's sound system, modernizing every single component. It was a true labor of love. The ultimate highlight of that farewell performance arrived when Eugene was right there at the helm, expertly operating the very sound system we had painstakingly pieced together. Witnessing his skill in that moment was an incredible full-circle experience, a testament to our shared passion, and a beautiful precursor to his future—because Eugene has since become a professional audio engineer.

Moving to Houston marked the beginning of a whole new chapter. I pretty quickly found my footing within a new church choir, which opened doors to many solo opportunities, often seamlessly integrated into the choir's selections. One performance that still gives me chills involved singing "Somebody's Praying" during a service honoring service men and women. The performance was accompanied by a video showcasing images of soldiers in the field, and it was made even more powerful by a young girl's heartfelt recitation of a letter to the troops. The combined effect was undeniable, leaving a lasting impression on all those present.

Looking back on it all, I feel this deep, deep gratitude for the solid musical foundation I was given. It wasn't just about learning notes or scales; it allowed me to make meaningful connections with people, gave me opportunities to lead in worship, and even sparked creativity in something as fun as children's ministry. This entire journey demonstrates that music training isn't just about mastering an instrument or hitting the right notes; It is about cultivating the ability to connect with hearts, inspire communities, and truly leave a lasting impact. Who knew those early scales and rehearsals could lead to such a fulfilling and enriching adventure? Definitely not me, but I'm sure glad they did!

Chapter Ten

Dreams Can Come True

Y ou know those mornings where everything just feels right? That was our last breakfast in Maryland before heading to Houston, TX. Pancakes, bacon, and the best company a dad could ask for: my wife and my son, Eugene. It felt like we were deliberately trying to memorize every bite, every laugh, before we embarked on this whole new chapter down in Texas.

Saying goodbye was, predictably, a tear-fest. You spend years watching your kid grow up, blossom into this amazing young man, and then suddenly, he's off forging his own path. It is a proud moment, for sure, but man, did it ache. Still, even with the lump in my throat, I knew this move was a must-do. My long-held dream of working with NASA? It was finally happening.

I had been working toward this for two decades. From a starry-eyed kid gazing up at the night sky, to busting my butt through academics, hitting a few walls, and taking on all sorts of different roles over the years, it all felt like one big, long prep session for this incredible opportunity. But as the car ate up the miles toward Texas, a little voice in my head kept nagging, "Are you really ready? Can all that stuff you've learned actually cut it at NASA?" It was a big leap, after all.

The drive became a total trip down memory lane, especially as I thought about all the awesome times Eugene and I had shared. I remembered the three months we had between my old job and starting at NASA. It felt like an eternity back then, but we packed it full. Golf driving ranges (he probably still out-drives me!), computer shows, and just... chilling. Looking back, those months vanished in the blink of an eye.

Memories from his childhood kept bubbling up, too, like our backyard football games. We were convinced that if we filled the football with helium, it would fly further. Yeah, probably a bit naive, but we kicked and threw with such hopeful abandon, exploring the limits of our homemade innovation. The 'impact' of that helium, real or totally imagined, created some seriously lasting memories.

But one experience truly stands out; it remains so vivid in my mind. When Eugene was a senior in high school, we got an unbelievable, almost surreal chance to play the mellophone with the University of Maryland marching band at a World Football League halftime show in Barcelona, Spain. The MD band director knew that my son and I played various brass instruments, including the mellophone, and invited us to join the UMD marching band for this event. The weeks leading up to it were a blur of rehearsals, the brass echoing, the drums steady—it was the soundtrack to our lives. And then, stepping off the plane, that warm, vibrant Barcelona air just hugged us. The city was absolutely buzzing, unlike anything we'd ever experienced.

Our days were a captivating mix of sightseeing and spectacle. We spent countless hours on bus rides, traversing the city and its outskirts, each journey offering new glimpses of Spain's diverse landscapes, from the bustling urban sprawl to the tranquil coastal vistas. It was on one of these rides, as we navigated the narrow, chaotic streets of the city center, that something truly unforgettable happened. There was this sudden, jarring thud. A delivery motorcycle, clearly in a desperate hurry, slammed right into the side of our tour bus! There was a collective gasp from the band, a moment of stunned silence, as we peered out to see the rider, remarkably unharmed but looking quite sheepish, quickly gathering his scattered packages and disappearing down the street. Our bus driver barely flinched. He just glanced in his mirror, gave a dismissive shrug, and continued on as if collisions were a daily occurrence in Barcelona. It was a bizarre, unexpected moment that only added to the surreal, adventurous nature of the trip.

Marching through the vibrant, sun-drenched streets of Barcelona, our mellophones gleaming under the Mediterranean sun, promoting the World Football League game alongside my son, was an unparalleled father-son bonding experience. We navigated ancient alleys and grand boulevards, the cheers of curious onlookers a constant backdrop to our synchronized steps. Sharing the anticipation, the exhaustion, and the shared thrill of performing on an international stage, culminating in the thunderous roar of the crowd during the halftime show, cemented this journey as the most profound and unforgettable experience of our lives.

Oh, and then there were all the autograph events we got to hit up. One particular memory shines the brightest: Johnny Unitas, the legendary quarterback for the Baltimore Colts, signing Eugene's football. But the story gets even better. Eugene had a cast on his wrist at the time, and Unitas, being the generous legend he was, signed *that* too! My wife, Lauretta, went into full protective mode, meticulously instructing the doctor to be extremely careful during the cast removal, just to make sure Unitas' autograph stayed perfectly intact. That cast, a truly cherished memento of a special moment, is still with me today.

As we continued our westward journey, rolling into Texas, the thought of contributing my skills and expertise to the future of space exploration fueled my determination. Yeah, there was still some uncertainty about what lay ahead, but I knew that this whole journey, with all its challenges and opportunities, had shaped me into the person I was meant to be.

We finally pulled into Clear Lake, Texas, a few days before my start date of July 3, 2000. I was so grateful for that extra time to just relax and get a feel for the area. Clear Lake, TX, instantly charmed us with its friendly locals and pretty scenery. We spent those precious days exploring local restaurants, taking leisurely walks along the waterfront, and just getting a vibe for our new home. It was a peaceful prelude to the demanding work schedule that awaited me, a chance to breathe and appreciate the moment before diving headfirst into the adventure of a lifetime. And as I looked forward, I knew I would always carry the memories of Eugene, of family, and of home right there with me in this brand-new chapter.

Building a Branch, Building Trust

Let me take you back to a very significant day for me at NASA's Johnson Space Center—July 3, 2000. That's when a brand-new chapter really kicked off, marking the start of this huge project: creating the Flight Hardware Development Branch, which we later affectionately knew as the Parts, Package, and Manufacturing Branch, or EV5 for short, tucked away in the Avionics Systems Division. Honestly, it felt monumental—not just figuring out what this new branch would *do*, but also where it fit into this massive, established organization. It was a bit like building a house from the ground up, requiring a clear vision and relentless dedication.

Getting the right people on board wasn't easy. I had to essentially 'poach' folks from other well-established branches, and that meant really earning their trust. Drawing on my past management experience, I made a conscious decision to go with a servant leadership style. My goal was to create an environment where everyone felt respected, and where I could provide clear direction without stifling all their brilliant individual initiatives. And let me tell you, that approach was vital. It really helped foster that buy-in and maximized the team's potential. We definitely aimed high, expecting top performance, but I always promised a kind of "safety net." I wanted the team to really push boundaries, knowing that if they put in an honest effort, I would have their back, even if things didn't always go perfectly.

Our branch's scope was quite expansive, touching all sorts of human spaceflight hardware development. We dealt with everything from EEE (Electrical, Electronic, and Electromechanical) part selection to radiation testing, electronic layout, and even full system design. All that broad responsibility meant we were constantly learning and adapting. To make things feel a bit more personal and really connect us to the mission, we started tying the astronauts' names directly to the items we built. It brought a real sense of pride to our contribution.

Speaking of astronauts, working directly with them was truly a unique and enriching experience. Just seeing firsthand how the technology we were building would actually be used in space was both exhilarating and, honestly, a little daunting. The astronauts themselves—their knowledge, their enthusiasm, their sheer dedication—were incredibly inspiring. It made the whole collaboration so rewarding. These interactions really hammered home the profound impact of our work, knowing we were helping these amazing individuals who bravely explore the unknown. And even with their esteemed positions, they always treated our team with such respect, which led to some unexpected and lasting friendships.

That's why the tragedy of the Columbia Space Shuttle disaster in February 2003 hit us so hard. Those close relationships we'd formed with the crew made the news of the shuttle's disintegration during reentry all the more devastating. It was a profound shock, a stark reminder of the inherent risks of space exploration and, well, the fragility of human life. The images of scattered debris? They were a somber testament to the sacrifices made in the pursuit of reaching beyond the stars.

After losing Columbia, the push to boost safety became paramount. That tragedy reinforced the unimaginable cost of overlooked vulnerabilities, making the prevention of future astronaut fatalities our top priority. A huge part of that new focus was identifying potential wing damage during ascent—a critical but often unseen hazard. Exploring a LIDAR-type camera to be attached to the shuttle arm for in-orbit wing inspections wasn't just some cool technical upgrade; it was a fundamental shift toward proactive damage assessment. This capability was crucial because even tiny damage, which was invisible from the ground, could turn catastrophic during re-entry. The significance of ensuring this system worked flawlessly can't be overstated. It echoed lessons we'd learned from decades of demanding aerospace engineering: new safety protocols, especially when human lives are on the line, needed exhaustive and rigorous testing. This dedication wasn't just about technical precision; it was about restoring public confidence in human spaceflight and upholding the integrity of the entire space program. Without the ability to detect and potentially fix such critical damage, crewed missions would just be too risky, jeopardizing both lives and the very future of space exploration.

Another challenge popped up with the Simplified Aid for EVA Rescue (SAFER) backpack. Drawing on some prior experience I had as a troubleshooter, we managed to pinpoint the cause of its failure and implement a corrective action. It was a good reminder that even the smallest components can have huge implications.

A significant portion of our branch's responsibility revolved around EEE parts—these are the fundamental electrical components in absolutely everything, from your smartphone to satellites. Understanding how they worked and interacted, especially in the harsh space environment, was crucial to their success. Ionizing radiation, a constant threat in space, can cause damage at the atomic level, potentially leading to degraded performance or even complete failure. So, rigorous radiation testing was totally essential.

That meant regular trips to the University of Indiana for high-energy proton beam testing became routine for us. One time, I kid you not, a squirrel decided to mess with the electrical system, shutting down the proton beam!

It was a wild reminder of how unpredictable even the most meticulously planned experiments can be. On another occasion, we brought laptops intended for use on the International Space Station (ISS) for testing, and they immediately failed under the proton beam. Talk about a wake-up call, which underscored the importance of thorough testing before anything makes it to space.

Our branch also maintained the crucial processes and standards for circuit card layout. This basically involved designing the physical placement and connections of components on printed circuit boards (PCBs) to ensure optimal performance and reliability. It is an intricate process that required careful consideration of signal integrity, thermal management, and manufacturability. Sounds complex, right? It was!

Over time, our team really coalesced into a super-efficient unit. Appointing Dave Beverly as Chief Engineer was a fantastic move, and fostering a close working relationship with him led to a strong bond, both professionally and personally. Through shared interests and mutual respect, I truly felt like I had found a kindred spirit, which totally solidified the foundation of a thriving and successful branch within NASA. Looking back, the journey of building EV5 wasn't just about creating a new department; it was about building trust, fostering innovation, and contributing to the ongoing exploration of the cosmos.

Oh, and our Branch wasn't just heads-down in the lab. We actively engaged in crucial agency and international discussions, bringing our unique perspectives on critical components and openly discussing both the hurdles we hit and the inventive solutions we devised. A standout moment for me was having the chance to unveil our cutting-edge EV5 EEE parts approach to the Japanese Space Agency (JAXA) in Tokyo. For me personally, this was a huge milestone: my very first presentation to a large, international assembly, with the added complexity of speaking through a translator. The prospect was, to put it mildly, nerve-wracking. It was a mixture of intense preparation and that thrilling fear of the unknown.

After the adrenaline of the presentation slowly wore off, I found some real peace in the quiet efficiency of our hotel. The hotel itself was a testament to quintessential Japanese hospitality—sleek, minimalist, but incredibly functional. Although compact, my room felt intentionally designed, with every amenity thoughtfully placed. The large window offered a mesmerizing view of Tokyo's sprawling, glittering cityscape, a stark contrast to the quiet calm inside. One evening, unwinding after a long day of meetings, a subtle tremor started. Initially, I thought it might be a huge truck passing by, but the rhythmic sway of the curtains and the gentle rattle of the mini-bar glasses quickly told me otherwise. It wasn't violent, more a prolonged, unsettling undulation that maybe lasted fifteen seconds. My initial surprise turned into a quiet fascination, then a slight unease as I realized the building was actually moving! A quick check of local news apps confirmed it: a minor earthquake, barely noticeable to seasoned Tokyoites, but certainly a memorable first for me. What a trip!

The rest of the trip went smoothly—a great blend of productive discussions and some much-needed cultural immersion. For the presentation itself, preparation was key. I methodically crafted my slides and practiced countless times. But the translator added a new layer of complexity. Maintaining a clear and deliberate pace became paramount, making sure the translator could accurately convey all the technical nuances of our approach. Despite the initial apprehension, the whole experience was incredibly rewarding. Seeing the audience's engagement, evidenced by nods and attentive expressions despite the language barrier, made all the effort worthwhile. The translator was a vital bridge, really facilitating a meaningful exchange of knowledge and fostering a collaborative spirit.

Beyond the professional aspects, the trip offered a cool glimpse into Japanese culture. The pre-meeting reception included my first-ever encounter with sushi. As someone who usually prefers a more traditional meal, this was definitely a step outside my comfort zone. But, armed with my training chopsticks, I gave it a try!

While not a new favorite, the experience broadened my perspectives and underscored the importance of embracing new experiences. The trip to Japan reinforced the importance of clear and concise communication and gave me a new appreciation for cultural differences in a global workspace.

Saying goodbye to Tokyo was a mix of emotions. There was definitely a sense of accomplishment bubbling up—we'd gotten so much done—but also a touch of melancholy. Leaving that city, with its incredible energy and all those unforgettable moments, felt like closing a really important chapter. From the super-demanding JAXA presentation that pushed me to my limits, to the tremulous rumble of an unexpected earthquake that reminded me of our planet's wild side, it really was a journey that stretched my professional boundaries. It also gave me a unique peek into how international teams collaborate, and frankly, just how unpredictable things can be.

But as one chapter closed, a new one was already well underway back home. We'd successfully gotten our new Branch up and running, fully operational and humming along. With that big step achieved, the next logical move was to find some serious backup. I knew I needed a Deputy Branch Chief, someone who wasn't just a warm body, but a true partner. I envisioned someone who could really share the load, tackle the complexities of running a successful branch, and basically have my back. This person would be key in overseeing daily operations, managing the team, and helping us navigate any challenges that popped up. And just as importantly, they'd be a valuable sounding board for all my wild ideas and a collaborative brain for problem-solving.

My hiring philosophy for this role was fairly straightforward: I wanted diverse perspectives. I'm a firm believer in the power of collective intelligence, and I often quote my own little mantra: "I am not smart, you are not smart, but collectively we are brilliant." My goal wasn't just to fill a seat, but to find someone with serious potential for growth, development, and ultimately, who could step into my shoes one day. And wouldn't you know it; I found exactly that in an incredibly competent individual who ticked every single box. I committed to mentoring her right then and there, preparing her to eventually take the reins as Branch Chief.

And the results? Well, they pretty much speak for themselves. Within three years, my Deputy was completely ready to take charge. This smooth transition allowed me to seize an amazing opportunity: accepting the role of Deputy Division Chief (acting on rotation) within the Avionics Systems Division (ASD) at the Johnson Space Center. ASD's work in ensuring our avionics systems are reliable is absolutely critical to our mission, and I was so eager to contribute at that higher level. This move felt like a pivotal step toward my long-term goal of eventually transferring to NASA Headquarters in Washington, D.C.

Looking back, this whole journey has been a whirlwind of continuous learning and growth. It has been marked by successful international collaboration, the strategic building of a strong team, and a deep commitment to fostering the next generation of leaders. The lessons I've picked up—both on the job and personally–shaped my contributions to the exciting field of space exploration.

Chapter Eleven

Transferring to NASA HQ

From Houston to Headquarters

January 14, 2004, marked a pivotal moment for space exploration. President Bush unveiled his Vision for Space Exploration, a bold initiative aimed at reinvigorating the United States' space program. NASA responded by creating the Constellation Program, a dedicated effort to bring that vision to life. This program was initially conceived at NASA Headquarters in Washington, D.C., and I found myself presented with a unique opportunity to be part of its genesis. Leaving behind my familiar surroundings at the Johnson Space Center in Houston, I prepared for a transfer to the nation's capital, eager to contribute to this ambitious endeavor.

Before the excitement of this career leap consumed me, I couldn't help but reflect on the events that had shaped the nation and the world in recent years. The memories of September 11th, 2001, were still vivid and unsettling. I recalled sitting on a plane, ready to take off from Houston to Los Angeles, when the pilot announced an unforeseen hold. Soon, the horrifying details of the attacks began to unfold. The news of the plane crash in Shanksville, Pennsylvania, resonated particularly strongly, reminding me of playing basketball against the Shanksville team back in high school. Even now, I want to extend my sincere gratitude to the brave rescue teams who risked their lives that day, embodying the resilience and spirit of America in the face of unimaginable adversity.

My time at the Johnson Space Center had been incredibly rewarding, allowing me to support a variety of initiatives that ultimately paved the way for my transfer to NASA Headquarters. I thrived on contributing to projects that were pushing the boundaries of space exploration. The opportunity to move to Washington, D.C. and take on a more senior role was a testament to the hard work and dedication I poured into my work at JSC.

Securing this position at NASA Headquarters was a game-changer for my career, and moving to Washington, D.C., felt like a huge step forward in my professional journey. Being based at headquarters opened numerous doors. I was able to connect with key NASA officials, gain invaluable insights into the inner workings of the agency on a much larger scale, and truly understand the nation's space exploration goals. While the transition wasn't always easy, navigating a new environment and a more complex organizational structure, it was undeniably worth every bit of effort. Though the Constellation Program Office eventually relocated to JSC, with oversight from the Exploration Systems Mission Directorate at NASA HQ, I remained in D.C., continuing to contribute to the agency's mission.

My journey, from supporting programs at JSC to being involved in the early stages of the Constellation Program at NASA Headquarters, has been a testament to the power of dedication, resilience, and the unwavering pursuit of a dream. It is a reminder that even in the face of adversity, we can strive toward ambitious goals and contribute to something larger than ourselves, pushing the boundaries of human knowledge and exploration.

Chapter Twelve

Another Devastating Event Struck the NASA Family

Working at NASA HQ in Washington, D.C., was always a dynamic experience. My office was usually buzzing with conversations, keyboards clacking away, and the NASA channel playing softly in the background. It was a pretty lively and focused atmosphere, even on a Friday afternoon.

One particular Friday in 2007, I was multitasking, as usual, glancing up at the screen now and then to catch the latest updates on our missions and launches. That's when something caught my eye—emergency vehicles swarming around Building 44 at Johnson Space Center in Houston. Building 44... that place hit close to home. It was where my office used to be when I worked at JSC. Instantly, my gut told me something was seriously wrong. I picked up the phone and reached out to one of my old bosses, Pat Pilola, to find out what was going on.

I can still vividly recall the moment Pat broke the news. There had been a shooting in the building. Almost everyone had been evacuated, but three people were still inside: Dave Beverly, my administrative assistant, and a support contractor we'd hired. The contractor, believing Dave was about to have him fired, had shot Dave. My administrative assistant was right there in the room. Dave tragically lost his life trying to protect her from the shooter, who then turned the gun on himself, right there in *my* old office. It was a heartbreaking and just unimaginable tragedy.

A few years before, Dave had asked for some extra help with his tasks, which is why we'd brought in that contractor. After hearing what happened, I couldn't help but replay the whole thing in my head, wondering if I had missed something during the interview process. Could I have done something, anything, to prevent it? The "what ifs" were overwhelming.

I wanted to give Linda, Dave's widow, some space to grieve, not wanting to overwhelm her with immediate calls. But the very next morning, I called her to check in and offer whatever support I could. Dave and I weren't just coworkers; we were best friends. Our conversation was a good one, full of shared memories—recalling our Saturday breakfasts with the motorcycle group and the many afternoons spent tinkering with our bikes in Dave's garage. I told her I would help in any way possible. Little did I know she'd actually take me up on it! She asked me to sing the song "Sheltered in the Arms of God" at two memorial services: one at the JSC main auditorium and the other at their church. I agreed, and we ended the call.

Now, Linda and I were part of the same congregation, so she'd heard me sing plenty of times. She'd witnessed numerous solos I had performed during various services over the years. Given how familiar she was with my voice and my regular contributions to worship, I naturally figured that was why she requested I sing at Dave's memorial service.

But I had a huge dilemma—I didn't know the song at all, and I wasn't even sure if a soundtrack was available. Plus, as a bass singer, I was worried it might be completely outside my vocal range. I explained the challenge to Lauretta, who immediately started calling Christian bookstores trying to track down the soundtrack. Meanwhile, I frantically packed for my trip from Washington, D.C., to Houston, TX. Luckily, she found one! So, I grabbed my portable CD player and headphones, and I literally learned the song on the 3-hour flight to Houston. Talk about a crash course!

Singing at my best friend's memorial service, in front of all my coworkers, and on live TV... that was a challenge I just *knew* I had to overcome. I sat in the green room getting ready for the service at JSC, periodically glancing out at the packed auditorium. TV cameras lined the aisles, and I was told the service would be broadcast live on the

NASA channel and one of the major internet channels. It was definitely a surreal experience.

Singing at my best friend's
memorial service

I remembered how I had managed to suppress my emotions when my sister's husband and daughter passed away, and I knew I had to dig deep inside myself for that same strength again. While I was on stage, memories of my time with Dave came flooding back. I shared a story with the audience about a time we went motorcycle riding, and I just couldn't keep up with him. I blamed it on my smaller ride, a Honda Pacific Coast 800 cc. So, Dave, being Dave, sold me his Suzuki GSX1100. But even with the faster bike, I still couldn't keep up with him. When I finally caught up, he looked at me with a perfectly straight face and said, "I don't think the issue is the bike." The audience actually laughed at that memory, and later, people told me they could really see Dave's humor shining through in that moment.

As I sang, my hand was shaking, not from nerves, but from the flood of emotions I was feeling. The lyrics brought back so many memories of the struggles we had faced together, the good times, and the bad. Looking into

Linda's eyes, I saw a mix of love, pain, and hope reflected back at me. It was a powerful moment that words truly couldn't describe.

A year after the memorial service, I was in Salt Lake City, Utah, for a meeting at a convention-type hotel. As I waited in line to check in, a lady approached me and asked if I was Mr. Yoder. I said yes, a bit curious about what she wanted. She told me she had seen the memorial service online and was so moved by my song. She said she had hoped to meet me one day just to say thank you. It was amazing, really, how far that broadcast reached.

My parents were eager to see the video recording of the memorial service, especially the part where I sang. As we watched it together, a quiet moment passed before my mother turned to me, her voice gentle but firm, and she said, "I want you to sing that song at my funeral." Years later, when she passed away, my father, with a mix of sorrow and memory in his eyes, gently reminded me of that specific conversation. He requested that I honor her wishes, and so, at her funeral, I stood and sang the song, a heartfelt tribute to her memory and a promise fulfilled.

Chapter Thirteen

The Sky's the Limit

Stepping into the role of Division Director for the Directorate Integration Office (DIO) at NASA Headquarters felt like hitting fast-forward into a whole new world. Honestly, it wasn't just some fancy promotion; it was the start of a truly transformative chapter in my life, marked by innovation, a collaborative spirit, and a plethora of unforgettable moments that have shaped who I am today.

Looking back, the whole journey to that position, and everything that came with it, feels almost surreal. I mean, one minute I was sketching out lunar exploration strategies on napkins during hurried lunch meetings, and the next I was on late-night calls with legendary astronauts like Buzz Aldrin. Every single experience wove itself into the rich tapestry of a story that's deeply personal to me. These weren't just tasks or responsibilities; they were incredible opportunities to contribute to something way bigger than myself, to actually push the boundaries of human knowledge, and hopefully, to inspire future generations.

And as I reflect on these incredible experiences, I'm struck by how much early encouragement and mentorship made a difference. I vividly remember my 5th-grade teacher, who somehow saw a spark of leadership potential within me. That seemingly small act of belief ignited a flame that would eventually guide my path, leading me to represent NASA in international dialogues, brief NASA advisory councils on critical topics, and actively contribute to some of the most ambitious space exploration endeavors of our time. Who knew, right?

The DIO role itself was a whirlwind. It demanded intense strategic thinking, meticulous coordination, and the ability to navigate complex challenges in a fast-paced environment. But beyond all the strategic meetings and complex briefings, it was the human element that resonated most deeply with me. The sheer passion and dedication of the entire team, their unwavering commitment to pushing the boundaries of what's possible, was truly inspiring.

Our Directorate was basically responsible for wrangling the Level 1 requirements for the Constellation Program. This was a *huge*, multi-center endeavor aimed at establishing a lunar base and planning future missions. Its complexity? Let's just say it presented a unique challenge for our IT department, requiring them to develop an integrated system that could capture requirements, manage design documentation, and track risks across the entire program. Talk about a puzzle!

One of the first hurdles I encountered was figuring out how to streamline our IT processes. I recall a particularly productive lunch meeting with the IT manager, where we really hammered out potential solutions. We ended up documenting the key points of one promising approach on a napkin, and then both of us signed it as if it were some official agreement. It was a testament to the collaborative spirit that permeated NASA—a willingness to explore innovative solutions even if they started with a humble napkin sketch.

My Lunar Obsession: The LAT

Beyond the technical intricacies, my time at the DIO was truly defined by my involvement with the Lunar Architecture Team (LAT). I started as a co-leader and then eventually led the team in developing strategies for future lunar surface operations. We dove deep into every aspect of lunar habitation, methodically evaluating mobility options, habitability designs, power generation strategies, protection measures against the harsh lunar environment, and optimal traversal routes. The sheer scope of the project was undoubtedly daunting, but the team's dedication and expertise were truly inspiring.

One evening, while working late, I received a phone call that I will never forget. It was Astronaut Buzz Aldrin, the second person to walk on the Moon. He had reviewed our draft recommendations and wanted to share his insights. We engaged in an extensive conversation, discussing various options and drawing parallels between his firsthand experiences on the lunar surface and our theoretical studies. To hear his perspective and connect with a literal piece of history was an incredibly humbling experience. I mean, how many people get to say that?

The responsibility extended beyond internal collaboration, as well. I was tasked with briefing the NASA Advisory Council on the progress of our study. This was particularly noteworthy because I had the opportunity to present our findings to the Honorable Harrison "Jack" Schmitt, the only geologist to have walked on the lunar surface as part of the Apollo 17 mission. His presence lent an unparalleled level of credibility to our work.

Our architecture team meetings themselves were often a fantastic blend of serious discussion and creative problem-solving. I fondly remember one particular occasion when I pointed out a flaw in a proposed concept. Pat, another team member, grabbed several Pepsi cans, rearranged the desk right there, and proceeded to illustrate a potential update to the architectural pattern. It wasn't the most conventional method, but it was surprisingly effective in visualizing the proposed solution! That's the kind of out-of-the-box thinking you get at NASA.

The LAT's architecture wasn't solely a NASA effort; it was a truly international endeavor involving contributions from space agencies worldwide. On a trip to the Netherlands for a collaborative meeting with a coworker, we encountered a funny, albeit slightly stressful, situation. As we were leaving the airport, I noticed a flash beside the road, dismissing it as nothing since I wasn't driving. About a mile later, we saw another flash. It turned out they were speed cameras, and we later received two speeding tickets upon our return to the States. That experience solidified my preference for not driving in foreign countries, fearing I would inadvertently break some obscure traffic law!

Our Agency was leading an international effort to develop a Framework for Exploration as part of the Lunar Architecture activity. We held a meeting in Germany with representatives from 13 international space agencies, fostering a rich exchange of ideas and potential contributions to the overall architecture. As the senior NASA official at the meeting, I led these discussions. That evening, sitting in the hotel, I reflected on the unlikely journey that had led me to this position. I was a small-town farm boy who had been told in high school that he would only be a farmer. Yet, here I was, leading international discussions on lunar exploration. It was a humbling and awe-inspiring moment—a real 'pinch-me' kind of realization.

The exposure of our work extended beyond the scientific community, too. I attended a meeting in Florida where we presented our architectural concepts. During a break, several TV stations, including the Disney Channel, conducted interviews with me. In preparation for future interviews and press conferences, I even underwent intensive media training, including mock interviews designed to test my ability to handle tough questions and potential "traps." By this point, media interviews had become second nature—who knew I had it in me?

As our LAT studies drew to a close, we shifted our focus to Constellation requirements verification. Then, the Constellation Division Director (CSD) announced his departure from the Agency. I was appointed to the role of CSD Director, marking another significant step in my journey at NASA. From napkin sketches to leading international collaborations, my time at the DIO was a testament to the power of collaboration, innovation, and a relentless pursuit of the dream of returning to the Moon. It was an absolute privilege to be a part of such a dedicated and talented team, working toward a common goal that truly inspired the world.

My Wild Ride as Constellation Systems Division Director

Stepping into the role of Director for the Constellation Systems Division (CSD) felt like a natural next step after my time leading the Directorate Integration Office (DIO). But "natural" doesn't quite capture how incredibly multifaceted and deeply engaging this job turned out to be. My brain was constantly buzzing, trying to wrap itself around both the mind-bending scientific challenges and the sheer logistical hurdles of getting humans back to the Moon and, eventually, setting our sights on Mars.

The Constellation Program itself was just... monumental. Seriously, a huge undertaking. The whole idea was to get humanity back to the Moon, not just for a quick visit, but for extended stays, and then use that as a springboard for further space exploration. At its heart, we were figuring out how to efficiently transport people and stuff to and from the Moon. This wasn't just about building rockets; it was about totally innovative stuff like extracting resources from lunar soil—regolith, as we called it—for building materials, habitats, and even better ways to get around up there. A massive goal was setting us up to live sustainably on the Moon while still being close enough to Earth for any emergencies. Plus, all our scientific goals were part of a larger global strategy, collaborating with space agencies worldwide. It was truly a team effort on an international scale.

My position meant I was often shoulder-to-shoulder with the Constellation Program Manager, giving direction and support straight from NASA Headquarters. Constellation was an incredibly intricate, integrated program, which meant we had to conduct vigorous design reviews at every single stage. We're talking requirements checks, progress updates, hitting key milestones—all absolutely crucial to ensure the whole system would click together exactly as planned. And get this: I had the unreal opportunity to work alongside aerospace legends like John Young. Yeah, THE John Young, test pilot, astronaut, commander of Apollo 16, and the ninth person to walk on the Moon! Hearing his firsthand accounts of what It is like on the lunar surface? That was just invaluable, a deeply enriching experience I'll never forget.

A large part of my responsibilities also involved overseeing the Commercial Crew Development Program, also known as CCDEV. This program was pivotal, key to helping the commercial space sector grow and flex its muscles. As Director, I was the source selection official for the CCDEV 1 contract. After reviewing numerous options and working closely with the evaluation team, I made the final selection decisions. And here's a fun fact about how serious this was: I had to immediately sign the selection documentation and hand it over to the legal team before even briefing the NASA Administrator or Deputy Administrator. That strict procedure was in place to guarantee that the decision-making process was totally impartial and free from any outside influence. No meddling allowed!

Managing both the NASA Constellation program and the commercially-driven CCDEV projects simultaneously felt a bit like juggling two very different kinds of balls. It sometimes felt like we were pursuing two distinct goals. But you know what? This dual approach proved to be extremely beneficial. It allowed us to develop and implement a much wider array of approaches to space travel. We weren't putting all our eggs in one basket, which, looking back, was a very smart move.

Before becoming Constellation Division Director, when I was the DIO Director, I co-chaired the Lunar Architecture Team (LAT), LAT-1, and then chaired LAT-2. These

were basically studies focused on what we could potentially do on the lunar surface. Later, as the Constellation
Division Director, we shifted gears to reviewing Mars
Design Reference Mission (DRM) studies. These efforts
focused on developing a "reference architecture"—basically a blueprint—for future missions to Mars.

The journey to Mars, though, that's a whole different beast. The increased distance amplifies every challenge: environmental factors, landing options, and the
sheer volume of supplies we'd need. I was involved in
developing Mars Design Reference Architecture 5 and
was honestly honored to be one of the four people who
formally approved its publication. Our discussions often
revolved around the optimal number of launches needed
to pre-position supplies on Mars before humans even
arrived, carefully weighing orbital dynamics and specific launch windows. Our absolute priority was ensuring
that pre-positioned supplies were safely on the Martian
surface and fully functional *before* we ever risked astronauts on the nine-month journey to Mars. The amount of
weight and sheer quantity of supplies required to sustain
humans on Mars while they waited for the return launch
window was just daunting. Verifying the safe arrival and
functionality of those pre-launched supplies was considered the best way to ensure astronaut safety. And it
was this approach, meticulously outlined in DRA 5, that
really served as our guiding reference for all future Mars
missions.

My time within the Constellation Program was marked by dedication, collaboration, and a relentless push to expand the frontiers of space exploration. It was a genuine honor to contribute to humankind's return to the Moon and to lay the groundwork for those future missions to Mars, helping to solidify our place as a multi-planetary species.

Looking back, being Division Director for both the NASA Constellation program and the commercially-driven CCDEV projects really did feel like juggling distinct objectives. But ultimately, this multifaceted approach contributed to a much more diverse and robust landscape for space travel, fostering the development and implementation of multiple ways to reach space.

As time went on, though, the Constellation Program started facing increasing financial and schedule pressures. It became clear that with a change in Presidential administration, the program's future, at least in its current form, was uncertain. So, when the Science Mission Deputy Associate Administrator offered me the opportunity to transfer to the Science Mission Directorate (SMD) as the Astrophysics Division Deputy Director, I jumped at the new challenge. Despite the shift, my time with the Constellation program and all the work we did to envision a future of human space exploration, both on the Moon and Mars, remains an incredibly meaningful chapter in my career. It was a privilege to contribute to such ambitious goals and to work alongside so many talented and dedicated individuals.

(Oh, and here's a picture of the plaque ESMD AA gave me for my contributions to the Constellation Program—pretty cool, right?)

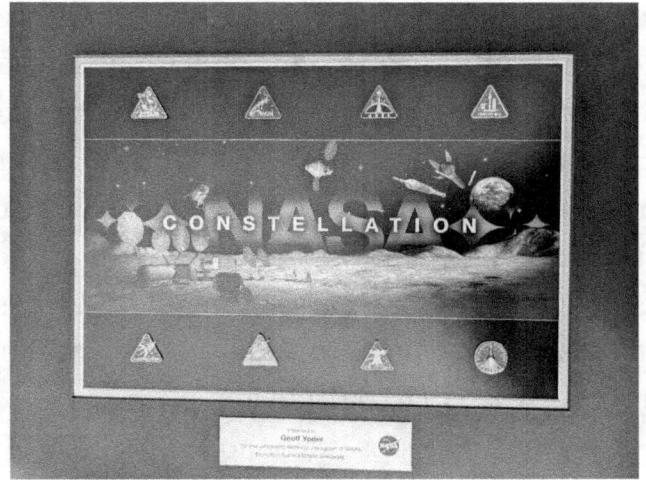

Constellation appreciation plaque

A Journey Through NASA's Astrophysics Division

The universe, it is just mind-blowingly vast and full of secrets, isn't it? Humanity has always been hooked on its mysteries. And at NASA's Astrophysics Division (APD), we're all about those big questions: How does this whole cosmic thing work? Where did we even come from? And the big one—are we alone out here? As the Deputy Director of the APD, I found myself right in the thick of this incredible quest, navigating a wild landscape of cutting-edge projects and discoveries that would make your head spin.

Now, my background? I was an engineer, which, I get it, seems a bit removed from the ethereal world of astrophysics. However, my job was all about bridging that gap, getting a real feel for how scientific ambition meets engineering reality. That meant spending at least 20% of my time chatting with scientists, engineers, and program executives, making sure we covered every single angle planning our missions. I was committed to ensuring everyone felt included and had a real sense of ownership. We even made a film showcasing the APD's key missions, featuring interviews with people across all departments, to really boost that shared pride and purpose.

The stuff the APD does? It is nothing short of revolutionary. Just think about the Hubble Space Telescope (HST). It is a cornerstone observatory, and it has completely rewritten our understanding of the cosmos. And its successor, the James Webb Space Telescope (JWST), is 100 times more powerful! It is pushing the boundaries of exploration even further. Hubble itself got a major upgrade, adding infrared (IR) capabilities, which let it peek through all that cosmic dust and gas, revealing hidden wonders we had never seen before.

Then there's the whole search for planets hanging out in the "habitable zone"—that "Goldilocks zone" where it is just right for liquid water, and maybe even life as we know it, to exist. Missions like the Kepler Space Telescope have been game-changers in this hunt. When I first joined the APD, we had only identified two celestial bodies that were potentially habitable. Now, that number is over 2,000!

The APD's commitment to research and technology development really paved the way for these mind-blowing discoveries, shaping the missions we're dreaming up for tomorrow.

Oh, and speaking of innovative stuff, let me tell you about the NASA Balloon Program. It is managed by the Wallops Flight Facility in Virginia, and it perfectly embodies that innovative spirit of the APD. We use these massive super-pressure balloons (SPBs) that can carry up to 8,000 pounds of scientific payload to altitudes over 100,000 feet!

(First photo: Credit NASA Balloon Program Office. Loading helium into the balloon.) These high-altitude missions are a super cost-effective way to gather crucial scientific data and are invaluable for testing new instruments and spacecraft technologies. (Second photo: expands to the size of a football stadium.) Credit NASA Balloon Program Office

In fact, Nobel Laureate Dr. John Mather even used the Balloon Program to validate some of his theories, which then led to the COBE satellite mission and, ultimately, his Nobel Prize. So, the Balloon Program isn't just a bunch of flights; It is a vital steppingstone toward some seriously groundbreaking discoveries.

However, even the most perfectly planned programs can hit a snag. In April 2010, we had a balloon launch in Alice Springs, Australia, that just went sideways. Strong winds caused the balloon to drift, and it ended up colliding with a parked car! This incident ultimately led to the grounding of the entire balloon program.

I flew out to the Wallops facility in Virginia to meet with the balloon team. As I sat in a room packed with balloon management and operators, you could just feel this thick cloud of nervousness hanging in the air. The silence was palpable, and I could tell they were all bracing themselves for a lecture about carelessness or incompetence.

My primary goal was to determine what happened without pointing fingers. I started the meeting by first asking a series of questions, but the answers I received were pretty vague. Everyone was still feeling super antsy. So, I did something a bit unusual: I had them all hold out their hands and then slap one hand with the other. I told them to consider their hand slapped, and that now we needed to focus on solving the problem, not playing the blame game.

After that, it was as if a switch was flipped. The team visibly relaxed, and we had a really good discussion about the issue and how to move forward. I later chatted about the experience with Dave Pierce, who was the Balloon Program Manager. He told me the team had been expecting a reprimand, so they were incredibly relieved that I was more interested in understanding the problems than in scolding them. It was a good lesson in leadership, I think.

Beyond all the science and engineering challenges, there were some very personal moments that shaped my journey. Longing for a proper break, my wife and I went on our first-ever cruise—it was a themed voyage featuring the Celtic Thunder Irish group. That trip sparked our love for cruising and gave us a rare chance to truly relax and recharge.

Even on vacation, my work was never far from my mind. I still proudly wore my APD SOFIA shirt, a constant connection to another remarkable project. SOFIA, by the way, was a modified 747SP aircraft, equipped with a 2.5-meter telescope. It lets scientists observe the universe from an incredible altitude of 45,000 feet, soaring above 99% of Earth's water vapor. Witnessing that fusion of aviation and scientific innovation aboard SOFIA was genuinely inspiring. The SOFIA Program Office even presented me with this cool plaque for my involvement with it!

SOFIA *plaque presented to me*

But life also brought some profound sadness. My older brother, just two years older, was diagnosed with glioblastoma, a devastating form of brain cancer. The prognosis was grim, and I spent his remaining months visiting him weekly, offering comfort and support. That experience was a stark, painful reminder of life's fragility and just how important compassion truly is.

Despite my engineering background, I was later appointed acting APD Division Director, a role typically filled by a scientist. I presented my vision to the National

Academy of Science and the NASA Astrophysics Advisory Committee, and I was delighted to receive their full support. It really felt like a testament to the power of open communication, strong collaboration, and a deep commitment to fairness across all disciplines.

In the end, my journey through the Astrophysics Division was a truly transformative experience. From overseeing groundbreaking missions to navigating unexpected challenges and even personal battles, I learned so much about the importance of teamwork, empathy, and this relentless, almost burning, pursuit of knowledge. The universe is still out there, beckoning, and the work of the APD remains absolutely crucial in unraveling its mysteries and answering those fundamental questions that, let's be honest, pretty much define our very existence.

From Red Wine Disaster to a Stellar Partnership

We all have those moments we'd rather forget. The ones that make us cringe, replay in our minds at 3 AM, and leave us wondering if we'll ever recover. For me, one such moment involved a transatlantic flight, a charming Parisian restaurant, and a rather unfortunate encounter with a carafe of red wine. But, as it turns out, sometimes those embarrassing moments can be the catalyst for something truly special.

My journey began with anticipation. I was heading to Paris to meet my European Space Agency (ESA) counterpart for the first time. We had spent countless hours on the phone, collaborating on joint projects, yet our connection remained purely professional, a voice on the other end of the line. As the plane soared over the Atlantic, I reflected on our past conversations and mentally prepared for the discussions ahead. We decided to meet for dinner the evening before our scheduled meeting, a chance to break the ice and get to know each other on a more personal level. Little did I know the ice would be broken in a way I never expected.

The restaurant was quintessential Parisian—cozy, intimate, with crisp white tablecloths and soft lighting. We were seated in a comfortable booth, engaging in casual conversation, getting past the formalities, and delving into shared interests and perspectives. Then, it happened. In a seemingly innocuous reach for a piece of crusty bread, my arm connected with a full glass of red wine. The crimson liquid erupted, painting the tablecloth with Jackson Pollock-Esque abandon, soaking my white shirt, and even reaching the wall beside me.

For a moment, time stood still. My face burned with embarrassment. But then, a strange calm washed over me. With a wry smile, I looked at my counterpart and quipped, "Sometimes my introductions are rather unusual."

The tension immediately dissolved. We both erupted in laughter. It was a shared moment of absurdity that transcended the awkwardness and forged an instant connection. The rest of the evening flowed with ease, the spilled wine becoming a running joke and a symbol of our newfound camaraderie. What started as a potential disaster transformed into the perfect foundation for a strong and productive partnership.

This lesson resonated even more deeply when I was tasked with leading the newly formed Office of Evaluation (OOE) at my agency. This involved a transition from my existing role within the APD team. The APD team was more than just colleagues; they were a close-knit family. Breaking the news that I was leaving and embarking on a new chapter was incredibly difficult. Saying goodbye to that support system filled me with a mixture of excitement and sadness.

However, the experience in Paris reminded me of the power of human connection. I knew that even though I was leaving the APD team, the relationships we had forged would endure. And I also knew that in my new role, the ability to build strong, trust-based relationships would be crucial to the success of the OOE.

So, the next time you find yourself in an embarrassing situation, remember the red wine in Paris. Embrace vulnerability, find the humor, and recognize that it might just be the catalyst for something truly extraordinary. After all, sometimes the most unexpected moments lead to the most rewarding connections.

A Journey Inside NASA's Quest for Improvement

You know NASA, right? The agency that sends rockets to space and explores faraway planets? They're always pushing the boundaries, not just in space, but also in how they run things. Every time a new administration comes in, it is like hitting a giant reset button, and they take a hard look at everything to figure out what could be better. One area of focus was how we conduct independent assessments—basically, checking on projects to make sure they're on track.

That's where the new Office of Evaluation (OOE) came in. This wasn't just another department; it was a big deal. The OOE was established to manage *all* independent assessments across the agency and report directly to the NASA Administrator. Talk about a direct line! And guess who they tapped to lead this new venture? Me! Which was kind of funny because my background wasn't in evaluation at all. I was actually a program execution guy.

It was definitely a curveball, but in a good way. It really reminded me of my old troubleshooting days back at Litton Systems. Before this, I was all about getting projects done, bringing them to life. Now, suddenly, I was in charge of, well, evaluating them. You might think, "That's weird!" However, I honestly think it was a smart move. I had been around long enough to witness firsthand where the old evaluation methods sometimes fell flat.

To be frank, I felt like some of those traditional independent assessments were mostly just "check the box" exercises. They didn't really unearth anything truly helpful for the agency. And that insight? That became my secret weapon. I wasn't just there to run an office; I was on a mission to transform the OOE from a bureaucratic necessity into something that actually helped NASA get better.

My core strategy was pretty simple: let's eliminate the rigid checklists and focus on assessments that actually add value. This wasn't about superficial compliance anymore. It was about digging into the real issues that impacted a project's success. My goal was to create a culture where assessments weren't seen as some scary threat, but as genuine opportunities to learn and grow. We needed open conversations and constructive criticism, not just ticking boxes.

Getting started was key. The first thing was to build a solid foundation for the OOE. I brought in a fantastic Deputy Director, and together, we developed a clear vision for what this office was going to accomplish. Our team wasted no time updating all the policies and procedures for both technical and programmatic assessments, making sure they reflected our new focus on adding real value. We wanted to ensure that our oversight processes were truly aligned with the bigger goal of keeping NASA's projects healthy and successful in the long run.

We were really hitting our stride. My Deputy Director was settling into what was a super challenging but incredibly rewarding role. And then, as often happens at NASA, another opportunity popped up—one that would put my program execution experience to quite a dramatic use. I got a call, and just like that, I was asked to take on the role of Program Director for the James Webb Space Telescope (JWST). Talk about a project of unparalleled ambition and complexity!

That unexpected shift really highlights the dynamic nature of leadership at NASA. While the OOE was poised to completely change how the agency evaluated its efforts, leading the JWST program provided an opportunity to apply those same ideas of rigorous evaluation and continuous improvement on a massive scale. So, NASA's journey toward better assessments was not just about tweaking processes. It was about building a culture of critical thinking, being highly adaptable, and relentlessly chasing after excellence at every level of the agency. And sometimes, that chase takes you in unexpected directions!

Navigating the James Webb Space Telescope Program

Imagine being part of something truly monumental, gazing into the universe's deepest secrets. That's the James Webb Space Telescope for you—a 6.5-meter cryogenic marvel, arguably humanity's most ambitious cosmic eye. However, for me, stepping into the role of its Program Director wasn't exactly a long-held dream or a smooth career climb. It was more like getting drafted into an elite but highly challenging special ops unit.

I was perfectly happy doing my own thing, to be honest. So, when the JWST Program Director role came knocking, not once but twice, I politely sidestepped it. I was aware of some of the issues bubbling under the surface, and frankly, I was content. But then came the call from the top brass himself, NASA Administrator General Charles Bolden Jr. That's when 'no' truly became 'yes.' His words still echo: "That's why I am directing you to take the position of JWST Program Director. Nobody else was telling me there may be problems." Well, when the Administrator of NASA directs you, you don't really refuse, do you? Challenge accepted.

So, there I was, suddenly at the helm of this incredible beast. My first priority? Just wrapping my head around it all. We're talking about a telescope unlike any other—groundbreaking science dictating a passively cooled cryogenic design, and the whole thing has to fold up perfectly to fit into a rocket. Oh, and did I mention the international partners? Understanding that complexity and who was responsible for what became my immediate mission. I decided early on: collaboration, not confrontation. My goal was clear: work with the fantastic project management team already in place. We all wanted the same thing—a successful launch, smooth operations, and the mind-blowing science that would follow.

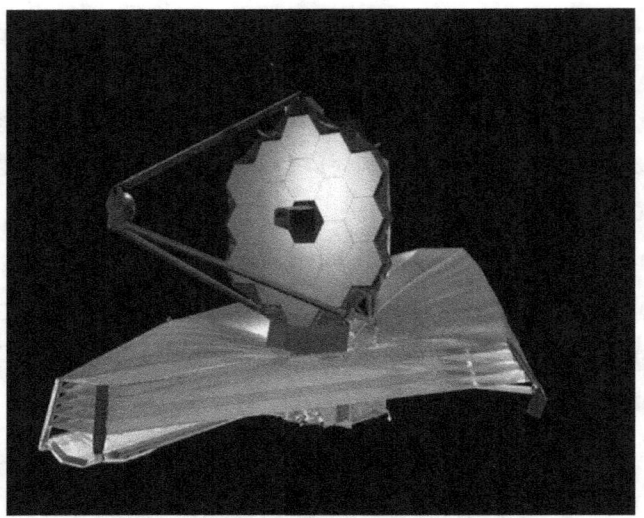

Artist image of JWST. Credit NASA

Regular progress reviews at Goddard Space Flight Center became our rhythm. But sometimes, you just had to get on a plane. I remember one particular whirlwind trip to Paris. I flew out in the evening, had a four-hour meeting with my European Space Agency (ESA) counterpart, and was back the next afternoon. It sounds wild, but those face-to-face discussions were absolutely crucial for building trust and keeping our international partnership humming.

Keeping everyone in the loop was another biggie. Working closely with my brilliant Deputy and Program Scientist, Dr. Eric Smith, we made sure stakeholders were well informed. That meant regular trips to Capitol Hill, where we discussed our progress, tackled any issues, and maintained transparency. But it wasn't all formal meetin gs...

Sometimes, these trips offered amazing opportunities to connect with the community. One time, Dr. Eric Smith and I teamed up with the incredible Astronaut Leland Melvin—both joined me via the Internet—to speak with over 300 middle school students in Pittsburgh. Our main message: no matter where you come from, you can contribute to something as epic as JWST.

The kids were totally hooked, especially when Astronaut Melvin showed off some science-in-space magic with a water bubble and an M&M. We talked about how vital science, math, engineering, and technology are for JWST, highlighting the telescope's ingenious folding mechanism for launch. To their delight, we even compared it to their beloved Transformer toys—which they totally got! Seeing their enthusiasm was such a powerful reminder of why we do what we do, inspiring the next generation in STEM. A local newspaper article even picked up on our discussions, which was pretty cool validation.

And then, there are the images. Among the countless breathtaking cosmic postcards JWST has sent back, one truly stands out for me: the Pillars of Creation in the Eagle Nebula. These towering structures of gas and dust, located about 6,500 light-years away, have fascinated astronomers ever since Hubble first spotted them in 1995. But JWST's shimmering, detailed portrait? It is providing scientists with unprecedented insights into how stars are born and how they shape their cosmic neighborhoods.

It is truly incredible to think I had a part in making that possible.

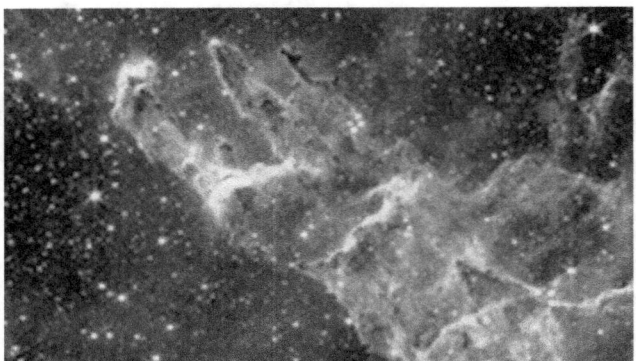

The Pillars of Creation, as seen by JWST.
Credit: NASA/ESA/CSA/STScI

But as anyone at NASA knows, careers are rarely linear! My JWST chapter, though incredibly rewarding, wasn't the final destination. Another "directed career change" was on the horizon, bringing new challenges and opportunities. Still, my journey with the James Webb Space Telescope remains a defining chapter in my professional life—a testament to the crucial role collaboration, dedication, and an unwavering pursuit of scientific discovery truly play. It was a wild ride, and I wouldn't trade it for anything.

Stepping Into the DAAP Role

My directed assignment was to step into the role of Deputy Associate Administrator for Programs (DAAP). My predecessor at DAAP had decided to retire, opening up this opportunity for me to gain a broader understanding of all the awesome work SMD was doing. Now, the DAAP gig definitely had some overlaps with the Deputy Associate Administrator (DAA) role. Still, my focus as DAAP was primarily on the technical side of our science missions, which was a perfect complement to the DAA's more programmatic focus.

Breaking the news to folks on Capitol Hill that I was moving on from JWST was... interesting. You could practically hear the collective gasp. One staffer was quite vocal about their concerns, saying it felt "irresponsible" to remove me from JWST leadership, especially given my no-nonsense, open communication about the project's progress and its inevitable bumps in the road. It was crucial to assure them that, as DAAP, I would still have an eye on JWST—just now alongside a bigger portfolio of, oh, only about 114 other missions! This explanation, while not exactly met with immediate high-fives, eventually seemed to land.

From day one, my goal was to foster a culture of genuine collaboration and proactive support. One of the first things I did was to visit successful bidders within three months of their project award. This wasn't about second-guessing decisions; those had already been made. It was just about showing up, showing support, and building a team vibe focused purely on mission success. This approach was a direct contrast to the intense selection process, which, of course, had to be super rigorous to pick the absolute best projects.

I'll never forget visiting one institution after their project received the green light. A seasoned team member, someone who'd clearly seen a lot, looked at me and said, "Thank you for visiting us. I have been here for 20 years, and you are the first senior person to visit us for a casual chat." That simple comment really hit home. It highlighted this kind of quiet gap in senior leadership interaction and underscored how important it is to connect with the very people who will make these ambitious projects happen.

Another key aspect of my leadership style is being accessible. I definitely had an open-door policy, but it wasn't so that every minor worry would land on my desk. Instead, it was an encouragement for staff to really think through their questions, formulate a solid approach, and then come talk to me. This led to much more meaningful and productive conversations, and ultimately, better outcomes. And I'm a firm believer in making decisions together. I would often "play devil's advocate," deliberately arguing against my own initial thoughts just to explore other angles and make sure we'd truly considered every perspective.

My move to the DAAP role was just one step on a continuing journey of leadership assignments within NASA. This entire endeavor was driven by a commitment to fostering collaboration, championing open communication, and making sure the brilliant folks within the Science Mission Directorate had all the support and resources they needed to keep pushing the boundaries of scientific discovery. As SMD continues to unravel the mysteries of the universe, it is this kind of dedication to excellence and proactive leadership that I believe will pave the way for all our future successes. Oh, and my reassignment to elevated positions continued when the SMD Deputy Associate Administrator left the Agency.

My Wild Ride as NASA's Deputy Associate Administrator

My role as DAA meant I was basically overseeing NASA's massive science portfolio in support of the Mission Directorate Associate Administrator. I was responsible for providing executive leadership and making sure all our missions—from monitoring Earth's climate to exploring deep space—ran smoothly. And honestly, my path to that role wasn't exactly a straight line. I remember thinking that it seemed like I was being reassigned every time someone left, which could make it appear that I couldn't hold down a job! However, those were all directed promotions, demonstrating that my management had considerable faith in what I could do.

As DAA, I focused on building a culture of respect and accountability within SMD. My management style was more about leadership than just plain managing. I would meet with my Division Directors weekly, giving them the freedom and responsibility to run their divisions within certain guidelines. It was all about the freedom to manage, but with accountability baked in. I truly believe in treating everyone on staff, from program managers and executives to scientists, engineers, support staff, and admin folks, with respect. Even if their roles sometimes seem more or less "important," everyone plays a vital part.

That commitment to respect really resonated with the team. I'll never forget one morning when an administrative assistant totally surprised me. She asked if I was a Christian. I was curious what made her ask, and she explained that she and the other admin assistants had been chatting. They all agreed I must be, because I treated them with the same respect I showed the scientists and engineers. They said they hadn't experienced that with other bosses. It was such a nice moment; it felt good to know I was "walking the talk" of trying to treat everyone with dignity.

Of course, the job wasn't all sunshine and rainbows. Holding people accountable for performance is a huge, sometimes challenging, part of being the DAA. And occasionally, those decisions lead to some pretty tricky situations. I remember one time when the MDAA and I had to remove someone from a key position for performance reasons. We followed all the legal steps, but the decision still resulted in a lawsuit. Sitting in that courtroom, explaining our decision, was incredibly stressful and tense. After four long hours on the stand, I was finally done. I drove straight to our cabin in West Virginia and, on a whim, decided to build a lower deck. I just needed to release some of that stress. Honestly, I might hold the record for the fastest deck construction ever!

Lower deck built after leaving the court room

Beyond managing my own team, I also played a significant role in collaborating with other agencies, like the National Oceanic and Atmospheric Administration (NOAA), to improve early weather predictions. And I worked very closely with NASA's Launch Service Provider (LSP) to make sure our scientific payloads launched safely and successfully.

Launching a science satellite or a rover is just a nerve-wracking experience, full of things that can go wrong. Sitting at those launch control monitoring stations, my mind would race, thinking about every single thing that might go sideways. I knew every step of the launch process, and I would literally hold my breath as the launch vehicle went through each phase. Things like MaxQ, when the rocket hits its maximum stress point, always made me a little nervous. Even though launches might look simple from the outside, they're actually pretty risky, and things could definitely go south.

Preparing for failure is actually a huge part of the process. For every launch, I prepared two speeches: one for a successful launch, and one for if something went wrong. Luckily, throughout my career at SMD, I only ever had to use the "We had a successful launch" speech. It was such a relief every time I got to deliver that good news to the press.

One memory that really stands out to me involved a joint celebration with ESA and NASA for the Rosetta probe landing on comet 67P/Churyumov-Gerasimenko. That mission was an engineering marvel, chasing, orbiting, and then landing on a comet! While the Rosetta orbiter kept doing its thing, emotions were super high as the lander's final communication approached. In a moment of stunned silence, I remember turning to Jim and comparing the experience of being in the room when my mother passed away. It wasn't the same, of course, but the attachment to that mission was incredibly strong, and its ending evoked some truly unexpected emotions.

My role as DAA at NASA's Science Mission Directorate was definitely demanding. It required a combination of leadership, technical know-how, and a lot of resilience. From overseeing complex missions to fostering collaboration and navigating challenging situations, I felt that I was playing a critical part in shaping our understanding of the universe and our place in it. It was truly an honor to help pave the way for groundbreaking discoveries that continue to inspire and amaze us all.

An Unexpected Interruption: A Wake-Up Call

Life has a way of throwing curveballs. Sometimes they're just gentle nudges, but other times, they're full-on collisions that make you rethink absolutely everything. For me, that collision came in August 2015, disguised as nagging back pain and a swollen leg.

I was at a two-day agency workshop, just brushing off the discomfort as a pinched nerve. But the next morning, the pain was excruciating. My wife, seeing me struggle, insisted I see a doctor. I'm usually pretty good at hiding pain, but this time, it was impossible. Our family doctor was out of the country, so we ended up at a 24-hour emergency care center. I confidently presented my pinched nerve theory, but the doctor, already looking suspicious, insisted on a thorough evaluation. The CAT scan and X-ray results hit me like a ton of bricks. Suddenly, I was being wheeled into a room, hooked up to machines, and given a shot in the stomach. It wasn't until I saw the look of sheer terror on my wife's face that the gravity of the situation truly began to sink in. The doctor explained that there were multiple clot issues. He asked which hospital I preferred, mentioning that an ambulance was ready. My mind struggled to process it. Ironically, my engineering background made me focus on the mechanics of the bumpy ambulance ride to Inova in Alexandria, VA, rather than the health crisis itself.

In the ICU, the doctors laid out the plan: I had a massive clot stretching from my hip to my ankle, and even more alarmingly, significant clots in both lungs. It was still difficult for me to fully grasp the extent of the danger I was in.

Amidst the sterile environment and medical procedures, a small act of kindness brought a much-needed ray of light. My five-year-old grandson visited, bringing a teddy bear adorned with the sweet message: "No day is so bad that a bear hug can't fix it." Pressing the bear's belly button triggered the song "So You Had a Bad Day." He would adjust my bed to make me comfortable, or maybe to ease his own anxieties. Either way, it temporarily took my mind off the looming threat of the blood clots. I still cherish that bear today, a constant reminder of life's fragility and the power of love. After three days and several procedures to address the clots in my leg, I was finally discharged. Two weeks of rest later, I was cleared to return to work part-time, gradually building back to full strength.

During my recovery, I was surprised by the relatively limited contact from work. My administrative assistant, bless her heart, had wisely instructed my staff to leave me undisturbed, coordinating with my wife to handle any truly urgent matters. This really showed the trust I had placed in my team and their ability to function autonomously. My leadership style had always been about empowering others, not micromanaging, and they totally proved their capabilities in my absence.

Even though I hadn't fully grasped the severity of the situation initially, seeing the worry etched on the faces of my loved ones really prompted a shift in my perspective. At 56, I started to seriously consider retiring earlier than the typical age of 65. Life's too short, you know?

My Unexpected Adventure Leading NASA Science

At NASA's Science Mission Directorate (SMD), we're like the conductors of this cosmic orchestra, making sure everything works together to guide groundbreaking missions from a scribbled idea on a napkin to a mind-blowing discovery among the stars.

Imagine getting the keys to a massive scientific spaceship, one that's exploring everything from our own backyard planet to the farthest corners of the cosmos. That's exactly where I found myself: at the very top of NASA's Science Mission Directorate (SMD) as the Acting SMD AA. This isn't just about managing budgets; It is about making game-changing decisions, navigating high-stakes launches, and constantly pushing for the next big scientific discovery. Think of it as NASA's ultimate science department, the one that steers the ship for all of America's civilian space science research, expanding the frontiers of four (the fifth discipline, Biological and Physical Sciences, added sometime after I retired from NASA) broad scientific pursuits: Earth Science, Planetary Science, Heliophysics, and Astrophysics. Basically, we're talking about everything from figuring out what makes Earth tick to exploring other planets, studying the sun, and peering deep into the universe. You can dive deeper into the NASA SMD Science Fleet right here at https://science.nasa.gov/science-missions/

Credit NASA SMD: Science Fleet

My journey to this acting role started pretty abruptly, as these things often do. When the SMD Associate Administrator announced his retirement, I was already serving as the Deputy. So, it was a natural next step for me to "step up to the plate" as the Acting Mission Directorate Associate Administrator.

The handover itself was symbolic. Dr. John Grunsfeld, a legendary astronaut who'd fixed the Hubble Space Telescope five times (seriously, five times!), literally handed me a figurative "key." He was all about the details and smart risks, and that really stuck with me as I faced my own challenges.

Symbolic handover of the keys to the Science Mission Directorate

When I took over, SMD was massive with an annual budget of around $5.6 billion and a whopping 114 missions, all in different stages—from just a twinkle in someone's eye to fully operational spacecraft zooming through space. My job was essentially to serve as the chief navigator, making sure all these diverse missions—from studying black holes and the sun to understanding our own planet and distant worlds—stayed on course. Every single one of them was built on past discoveries and aligned with the big-picture recommendations from the National Academy of Sciences, which is a pretty big deal.

I often get asked what It is like being an engineer during my time as Acting Science Mission Directorate Associate Administrator, when this position is typically filled with a scientist. People seem curious about whether there might be some inherent conflict between engineering and scientific pursuits.

Honestly, for me, it was seamless. My inquisitive nature for understanding and participating in exploration for both human and robotic exploration always fascinated me. My leadership style is to understand the options, challenges, and opportunities to gain knowledge, both for scientific and technical advancements, with a goal of increasing our curiosity about space and, at the same time, improving and advancing capabilities on Earth to help humankind. This is not in conflict with overseeing an organization as an engineer but rather merging the best of both worlds.

I have always believed that maintaining the highest level of integrity and treating everyone with respect, just as I want to be treated, should be the standard, regardless of the kind of work I am doing or what organization I am leading. That was something I constantly strived for. We had important missions to implement, and keeping focused on the task at hand was paramount.

The sheer exhilaration of overseeing the Science Mission Directorate is almost indescribable, a privileged vantage point from which to witness humanity's relentless pursuit of understanding a universe so vast and so utterly enigmatic, beckoning us to unravel its secrets.

A big part of my role involved working with the National Academy of Science, which, at the direction of Congress, conducts Decadal Surveys for each scientific discipline—Earth Science, Heliophysics, Planetary Science, and Astrophysics. These surveys, conducted every ten years, are comprehensive studies that gather input from scientists and engineers, along with programmatic insights, from around the globe to recommend priorities for each science area. The process involves evaluating different concepts and the underlying theories, which helps determine what should be the nation's focus. These surveys also provide guidance on how to prioritize options when funding isn't sufficient for all missions.

In my view, the Academy's Decadal Surveys served as the governing roadmap for the Science Mission Directorate. The NASA Advisory Council for the Mission Directorate and the Science Divisions offered further guidance. As a trained engineer and program manager, I was not focused on a specific scientific discipline, such as Astrophysics or Heliophysics, but rather had an appreciation for all the science disciplines in the Mission Directorate portfolio. I relied on the Division Scientists to provide me with key scientific information beyond what was stated in the Decadal Survey so that I could balance priorities across the Mission Directorate. I often received comments from scientists that they appreciated my giving equal attention to all the science disciplines and not being biased toward a specific discipline. I was honored to be part of NASA's exciting science missions and enjoyed working on various technical and scientific endeavors.

Take NASA's **Earth Science Division (ESD),** for example. Their missions are all about understanding our planet's interconnected systems, from a global scale down to the smallest processes. ESD provides technology, expertise, global observations, and applications that help us understand the myriad connections between our planet's vital processes and the climate effects of both natural and human-caused changes.

All of these efforts are incredibly important for understanding our planet and ultimately helping humanity. We've seen real-world impacts, like the early tsunami warning in the devastating 2011 Japanese Tsunami. US Ambassador to Japan Caroline Kennedy stated that the NASA/JAXA mission was credited with saving over 1,000 lives thanks to its early detection and warning.

Another great example is the NASA Cyclone Global Navigation Satellite System (CYGNSS), which utilizes eight micro-satellites in a unique way to measure wind speeds over the Earth's oceans. This helps us to better understand and predict hurricanes. For more information, check out .

One of my personal favorite missions is the Soil Moisture Active Passive (SMAP) mission. This satellite orbiting observatory measures the surface soil water content anywhere on Earth. This data is invaluable for providing information for irrigation services, identifying dry conditions that are ripe for fires, and gathering other key scientific data that helps us better understand our Earth.

There are so many Earth Science missions, all working to help us better understand our Earth and its ever-changing conditions. I view it as a way to better appreciate, understand, and care for the incredible planet we live on.

Then there is the **Heliophysics Division** that helps us unveil the Sun's influence on Earth and beyond. The sun, a seemingly constant presence in our lives, is far more dynamic and influential than we often realize.

Space is not an empty void, but rather an extension of the sun's atmosphere. The sun constantly emits a stream of particles and energy known as the solar wind, which carries the sun's magnetic field throughout the solar system. This dynamic solar atmosphere surrounds everything from the sun itself to Earth, the planets, and extends far beyond, shaping the environment in which we live.

Understanding this interconnected system is crucial for several reasons. First, it expands our fundamental knowledge of how the universe works. Second, and perhaps more practically, it helps us protect our technology and astronauts in space from the potentially disruptive effects of space weather. Solar flares and coronal mass ejections can interfere with our communications, damage satellites, and possibly even disrupt power grids on Earth.

Each mission occupies a carefully planned vantage point, designed to observe and understand the flow of energy and particles throughout the solar system. This coordinated effort enables scientists to piece together a comprehensive picture of the sun's influence on the heliosphere.

I also got to keep a close eye on the development of the Solar Probe Plus, which later became the awesome Parker Solar Probe. For many, the Probe embodies the cutting edge of solar exploration. This remarkable space-craft travels at speeds exceeding 430,000 miles per hour, so fast that it could travel from Philadelphia to Washington, D.C. in just one second! Then there is the insane heat shield, designed to survive temperatures up to 2,500 degrees Fahrenheit while traveling within 3.8 million miles of the Sun.

The importance of understanding space weather extends beyond protecting spacecraft and astronauts. The Earth itself is vulnerable to the sun's activity. This was made clear during testing of laptop computers for use on the International Space Station at the University of Indiana's high-energy proton cyclotron facility. One set of laptops failed almost immediately after the high-energy proton beam was activated, effectively mimicking the effects of space weather. The hard drive and monitor were rendered useless. Interestingly, another company was also present, testing their own computers not for space travel, but to identify the cause of computer failures at higher elevations on Earth. The hypothesis was that space weather events were causing computers to lock up or reboot, requiring modifications to the computer design to account for these solar influences.

So, you see, the study of how the sun affects us on Earth and the larger solar system helps us provide for better understanding of the world we live in and requires both scientific and engineering advancements.

The **Planetary Science Division** fosters our innate curiosity and the exploration of our solar system. From stargazing as a child to following the latest space mission updates, a deep-seated curiosity about the universe seems to be woven into the fabric of humanity. The desire to understand our place in the cosmos, particularly within our own solar system, fuels countless hours of research, engineering innovation, and exploration. The Planetary Science Division embodies this spirit, pushing the boundaries of what's possible in spacecraft design and robotic operation to unlock the secrets held within our celestial neighborhood.

For decades, NASA has been at the forefront of planetary exploration, sending probes and rovers to every planet and numerous smaller bodies within our solar system. This relentless pursuit of knowledge isn't just about satisfying our curiosity; It is about understanding the conditions that could support life—both on Earth and beyond.

One of the most captivating ongoing missions is the Perseverance rover's exploration of Mars. This marvel of engineering, equipped with a suite of advanced instruments and even a demonstration of helicopter ingenuity, has revolutionized our understanding of the Red Planet.

The sheer complexity of designing, landing, and operating a rover the size of Perseverance on Mars is truly remarkable. The thin Martian atmosphere presents unique challenges, rendering traditional parachutes inadequate for a safe landing. Engineers had to devise innovative solutions, like a "sky crane" that delicately lowered the rover to the surface using a combination of thrusters and strong cables. This audacious feat of engineering underscores the ingenuity and dedication required to push the boundaries of planetary exploration, reminding us that the pursuit of knowledge can drive technological advancements that benefit all of humanity.

The successful entry of the JUNO spacecraft into Jupiter's orbit was one of those moments that brought profound relief and excitement. The Washington Post even wrote about how the successful maneuver was a testament to meticulous planning. I remember sitting at one of the monitoring panels and turning to the Principal Investigator, and all I could manage was a speechless, "Wow, now for the real science." Pretty cool, right?

Ultimately, the exploration of our solar system is more than just a scientific endeavor; It is a reflection of our innate human desire to understand the universe and our place within it.

The universe, in its vastness and complexity, holds endless mysteries. As **NASA's Astrophysics Division** aptly states, it encompasses everything—space, matter, energy, time, and indeed, us. This division dedicates itself to unraveling these mysteries, pushing the boundaries of our knowledge, and sharing these discoveries with the world. NASA's astrophysics missions, guided by the Astrophysics Decadal Survey, strive to expand our understanding of the cosmos.

From the iconic Hubble Space Telescope (HST) to the groundbreaking James Webb Space Telescope (JWST), space observatories have revolutionized our understanding. HST's observations of thousands of distant stars and the confirmation of supermassive black holes at the centers of most galaxies, including our own Milky Way, have fundamentally altered our perspective. The 2010 Astrophysics Decadal Survey further highlighted the importance of understanding dark energy, the mysterious force driving the accelerated expansion of the universe. Missions like the European Space Agency's EUCLID and NASA's Roman Space Telescope are designed to shed light on this enigmatic phenomenon.

The history of astrophysics is a testament to the evolving nature of scientific understanding. What was once considered absolute truth can be challenged and redefined by new evidence. The geocentric model, adapted by astrophysicists in the second century and placing Earth at the center of the universe with the sun revolving around it, was eventually replaced by the heliocentric model championed by Nicolaus Copernicus in the 16th century, where the Earth and planets revolve around the sun. A striking example of this pursuit of knowledge is the story of Bob Williams, a Principal Investigator who used his allocated HST observation time to capture the "Hubble Deep Field." Against skepticism, he pointed the telescope at a seemingly empty patch of sky for ten days, resulting in a groundbreaking image revealing a multitude of galaxies previously unseen.

While ground-based telescopes remain valuable, space-based observatories allow us to overcome atmospheric interference, providing unprecedented clarity. NASA's investment in missions like KEPLER, SPITZER, NUSTAR, HST, and JWST demonstrates their commitment to pushing the frontiers of astronomical observation. JWST, in particular, has exceeded expectations, delivering images and data that may require significant revisions to existing textbooks. Its advanced technology, capable of operating under extreme temperature conditions, allows it to observe the universe in ways never imagined before.

Of course, there were extensive stakeholder interactions, ranging from detailed briefings with White House advisors to intricate negotiations with congressional staff and direct discussions with members across both chambers. These engagements were absolutely critical, serving as the essential conduit for conveying the nuances of our objectives, garnering support, and collaboratively shaping the legislative and policy landscapes required to successfully advance our missions and achieve our strategic goals, and relied on my program management skills to weave the information together into a seamless narrative.

Then there was the Mars Atmosphere and Volatile EvolutioN (MAVEN) mission. This spacecraft was designed to determine why Mars lost its atmosphere, a crucial question in our understanding of the planet's history and potential for life. Its launch window was closing fast. Miss it, and we'd be looking at an 18-month delay and a significant increase in costs. Everything was going smoothly until... yep, a government shutdown hit. MAVEN, being seen as 'non-critical' (ouch!), suddenly got a stand-down order. This put the whole launch in jeopardy.

But we got creative. We realized MAVEN could act as a crucial backup communication relay for our Mars rovers. Those existing relays were getting pretty old, and if they failed, we could lose contact with our rovers on Mars—mission over! By arguing that MAVEN was essential for keeping *those* missions alive, we could get it reclassified as critical. This took some serious behind-the-scenes coordination with various NASA offices, the White House, and more. But guess what? We pulled it off! MAVEN launched right on schedule, a real testament to thinking outside the box, sheer grit, and the unwavering teamwork of everyone involved.

Throughout all this, communication was key. I was constantly briefing presidential advisors and members of Congress, as well as their staff, on Capitol Hill. It was all about keeping everyone informed and involved in the process.

One meeting that really sticks out was with Representative John Culberson. We were talking about the Mars rover and its little helicopter sidekick. We hammered out a crucial agreement: the helicopter would be called a 'technology demonstration,' not a critical part of the main mission. This way, if the chopper decided to quit early, the whole rover mission wouldn't be labeled a failure. He totally got it, and we made sure that it was spelled out in NASA's directives.

Of course, not every mission makes it to the finish line. I had to visit Capitol Hill multiple times to explain why we had to cancel a mission, trying to justify those tough decisions in the face of tight budgets. I even put it to the staffers plainly: "Should we focus on keeping costs down like we've been told, or should we just let missions go over budget and move on?" It is a tough tightrope walk, balancing big science dreams with fiscal reality.

So, you see, my time as Acting Science Mission Directorate Associate Administrator was more than just a title; it was an odyssey brimming with excitement, wonder, and the thrilling adventure of pushing the very boundaries of human knowledge both scientifically and through engineering breakthroughs. Every day presented a fresh canvas for imagination, challenging us to envision the next frontier of discovery. But beyond the exhilaration, the profound purpose of the role lay in its unwavering commitment to humankind—masterminding the intricate dance where the rigorous precision of engineering met with the boundless curiosity of scientific inquiry. Our mandate was to forge pathways that allowed these two critical disciplines to intersect, not as disparate fields, but as symbiotic partners, ensuring that the theoretical breakthroughs of science found their practical, innovative application through engineering, and that engineering advancements enabled ever more ambitious scientific pursuits. The ultimate goal was to translate complex theoretical concepts into tangible, mission-driven realities, and vice versa, in a manner so succinct and seamless that the transition from blueprint to breakthrough felt utterly natural and inevitable. This was truly a journey dedicated to expanding the horizons of human understanding and capability.

Chapter Fourteen

Reflections on a Final NASA Launch

For many, witnessing a rocket launch is a lifelong dream. For me, it was a professional privilege, a testament to years of dedication and hard work. But one launch, the OSIRIS-REx mission to scoop up samples from the asteroid Bennu, held a particularly poignant significance: it was my last as a NASA employee. It was a moment that carried the weight of my entire career, a fitting end to a journey filled with excitement and challenges.

The OSIRIS-REx mission, a seriously bold endeavor to bring back a piece of an asteroid, perfectly encapsulated the spirit of NASA. This mission perfectly exemplified our nation's quest to boldly go and study our solar system and beyond, all the better to understand the universe and our place in it. Honestly, NASA science is the greatest engine of scientific discovery on the planet, and OSIRIS-REx truly embodied our directorate's goal to innovate, explore, discover, and inspire. It was the perfect mission to cap off my career.

Standing in front of the rocket launching
OSIRIS-REx

That day was filled with a mix of emotions. Sharing my final launch with my family and friends as VVIP guests added a truly personal touch to the whole momentous occasion. It also highlighted the bittersweet nature of my departure. After delivering my concluding overview at the OSPII building at the Kennedy Space Center, I made my way to the launch control site for one last time. That familiar journey from OSBII to the control center sparked a wave of nostalgia, a reminder of countless previous launches and all the dedication required to make them a reality.

As the countdown neared its end, a group of us went up to the rooftop, eager to witness the spectacle first-hand. With just 30 seconds before liftoff, the tension was palpable. Then, right at zero, the flames ignited, lighting up the sky with an almost blinding light. The powerful rumble of the engines filled the air, followed by the awe-inspiring ascent of the rocket. Moments later, those incredible reverberating soundwaves washed over us, a visceral reminder of the immense power propelling the mission. It was a sensation that honestly never lost its impact, a true testament to the precision and dedication required for space exploration.

Watching that rocket climb into the heavens, I couldn't help but reflect on the countless hours of preparation, the unwavering teamwork, and the years of dedication that had all culminated in this single moment. Each launch, including this final one, represented the pinnacle of human ingenuity and collaboration, a true testament to what can be achieved when individuals unite under a common goal.

As the rocket disappeared into the distance, a profound sense of accomplishment and gratitude just washed over me. This final launch wasn't simply an ending; it was a celebration of a career filled with challenges overcome, triumphs achieved, and memories that would truly last a lifetime. The unwavering support of family, friends, and colleagues only amplified the significance of the moment.

I often thought about the impacts of NASA missions beyond the mission itself. It is remarkable how much of our everyday lives have been touched by something as far-out as space exploration. From the comfy mattress I sleep on to the sturdy buildings we live and work in, I've noticed how NASA's "spinoff" technologies have quietly woven themselves into the very fabric of our daily existence. It really just goes to show how investing in pushing boundaries in space can lead to some seriously amazing and often unexpected benefits right here on Earth. And honestly, I bet that as we keep reaching for the stars, we're going to see even more breakthroughs that make our lives safer, healthier, and just plain more comfortable.

From the precision of GPS to the efficient management of water resources, to new measurement techniques used for LASIK eye improvements, and countless other benefits. NASA's journey to the cosmos is enriching life on Earth in profound and often unseen ways. It is an incredible thing to have been a part of.

One of the most significant examples of this impact, in my opinion, is GPS. I remember when it was just a thing for fancy cars, but now? It is everywhere! Originally, this technology was developed to make satellite navigation more precise for space missions. But fast-forward to today, and GPS correction technology is virtually indispensable. Beyond just telling me how to get to the grocery store, I've learned it is also revolutionizing things like farming—farmers use it to perfectly water and fertilize their crops, boosting yields and being kinder to the environment. And get this: It is a huge piece of the puzzle for self-driving cars, which will change how we travel and move goods around. Pretty neat, right?

I also find it fascinating how the search for water on other planets has actually benefited us here on Earth. All those clever techniques NASA developed to find water sources way out in space? They're now being used to pinpoint and manage our precious water resources right here on Earth. This is particularly important, especially in areas with limited water sources, as it helps us manage water more efficiently and sustainably. It is like they're looking for alien water, and we get better taps!

And speaking of data, imagine trying to make sense of the enormous amount of information that comes back from space missions. It is mind-boggling! That challenge actually spurred the creation of some amazing visualization tools. The NASA Visualization Explorer is a prime example. It is such a powerful tool because it takes really complex scientific data and turns it into visuals that are easy to understand. This not only helps researchers dig deeper into their discoveries, but it also enables regular folks like me to grasp complex scientific concepts. It is awesome for discovery, and it empowers educators and the public to engage meaningfully with scientific data.

Even something as niche as powdered lubricants developed to handle the extreme conditions of space has found its way back to us. These specialized lubricants are now being used in various industries to reduce friction, enhance machine efficiency, and extend their lifespan. Who knew space dust could do so much good?

But NASA's contributions aren't just about cool gadgets and tech. What I really appreciate is how they champion open access and collaboration. They actively promote sharing their innovations through resources like the Software Catalog. This massive library is packed with software tools and technologies originally developed for space exploration. By making these resources available to entrepreneurs, researchers, and developers, NASA fosters a spirit of innovation and empowers the next generation to build on its incredible legacy. It is like they're saying, "Here's our cool stuff, go make more cool stuff!"

So, NASA's impact extends far beyond just sending rockets into space. Through its commitment to innovation, collaboration, and sharing knowledge, NASA has developed technologies that are truly making our world better. As we continue to explore the universe, I'm genuinely excited to see even more groundbreaking innovations emerge, constantly blurring the lines between exploring the cosmos and making life better right here on Earth.

Ultimately, every launch marks a new chapter in the ongoing narrative of space exploration. This final chapter of my career with NASA served as a fitting tribute to the adventure and wonder of spaceflight, a journey that I'll forever cherish. My final launch was more than a good-bye; it was a celebration of a life dedicated to pushing the boundaries of human knowledge and exploration, a legacy that will continue to inspire generations to come.

Chapter Fifteen

A Retirement Celebrated—My Farewell from NASA

So, after what feels like a lifetime dedicated to figuring out how to get things into space and back, my amazing career at NASA finally wrapped up. But let me tell you, retirement wasn't just about saying goodbye; it was a full-blown celebration of a journey built on service, integrity, and, most importantly, building incredible teams. The send-off? It was truly mind-blowing, a real testament to the idea that one person can actually make a difference, not just on those groundbreaking missions, but on the awesome folks who make them happen.

The tone for the whole bash was set right away by General Charles F. Bolden Jr., our former NASA Administrator's, comments to the press. He said something that really resonated with me, acknowledging my contributions. "With more than 16 years in the industry and 16 years at NASA, Geoff's story is not only of individual success and hard work, but also of NASA's transition to a new era of space exploration, in which he played many key leadership roles. He has accomplished what most of us come here hoping to do—move our mission—and America's space program—forward." Hearing that from him? Pretty humbling if I'm being honest.

Surrounded by my incredible wife and son, the party focused on so much more than just the successful missions we pulled off. Speaker after speaker kept bringing up the high ethical standards I tried to uphold throughout my career, especially how I always made an effort to treat every single team member with respect and care, regardless of their role. It was a powerful reminder that true leadership isn't just about the big breakthroughs, but about how you treat the people who make them possible.

Then came the tokens of appreciation, each one more meaningful than the next. My colleagues at NOAA gave me something truly fascinating: a picture of the weather map from the very day I was born in Meyersdale, PA, complete with a detailed account of the weather conditions that day. It was such a thoughtful gesture and a cool reminder of all the collaborative efforts we shared to improve our nation's weather forecasting.

Perhaps the most cherished gift of all was a "Mission Portfolio" book. This thing was packed with comments from scientists, engineers, administrative staff, and so many others, documenting the diverse missions I had been involved with throughout my career. Flipping through it, reading all those personal anecdotes and acknowledgements, was a powerful reminder of the incredible collaborative spirit that truly drives NASA's success.

Mission Portfolio book

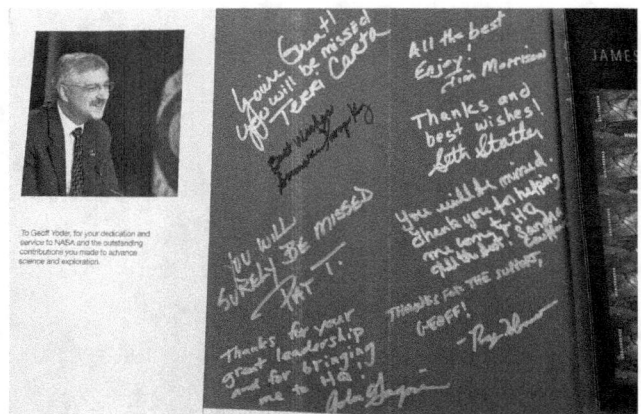

Mission Portfolio book

To further commemorate the occasion, I was presented a signed plaque from Astronaut and Kennedy Space Center Director Robert Cabana featuring a flag that had actually flown on Space Shuttle STS-135. This gift, symbolizing all the shared experiences and collaborations with the amazing team at the Kennedy Space Center, meant so much to me. Imagine that flag traveling more than 5.2 million miles and completing 200 orbits of the Earth! Other plaques arrived from the Johnson Space Center and various NASA Centers, but two really stood out. Two from NASA Administrator General Charles Bolden Jr. featured all the mission pins associated with the Space Shuttle, from its very beginning to its retirement. The other commemorated the missions to the International Space Station (ISS) from its inception up until the day I retired. These weren't just plaques; they were incredible symbols of years of dedication and commitment to pushing the boundaries of space exploration.

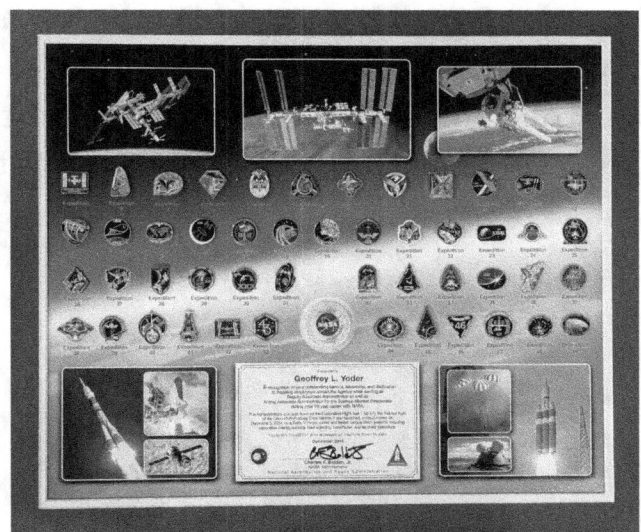

ISS Mission Pins

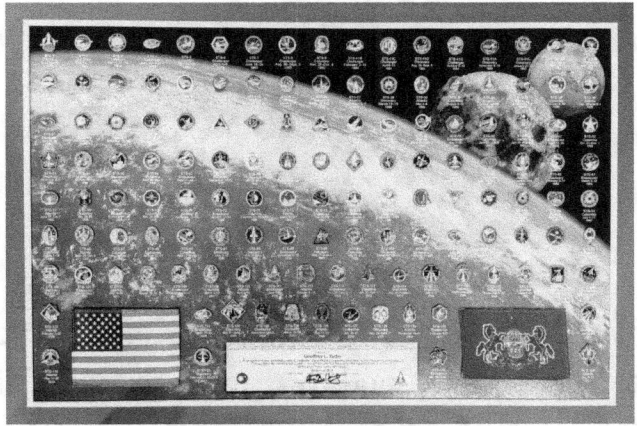

Shuttle Mission Pins

A particularly touching gesture came from Congressman Chris Van Hollen, who actually flew a United States flag over the U.S. Capitol in honor of my service to NASA and the country. While Congressman Van Hollen was not physically at the meeting, this really hit me—a poignant recognition of the dedication that defined my time there.

Finally, in a perfectly whimsical and humorous gesture, the Astrophysics Division created a bobblehead doll of me as a parting gift. A little goofy, maybe, but a great reminder that sometimes, it is the little things that really make a difference.

My bobble head

This retirement celebration was so much more than just a farewell; it was a powerful tribute to a career that, in retrospect, was defined by leadership, integrity, and a genuine commitment to the incredible people and the awe-inspiring mission of NASA. The gifts, the speeches, and all the shared laughter served as a testament to the profound impact one individual can have, leaving behind a legacy that I hope will continue to inspire for years to come.

Chapter Sixteen

Retirement: A New Launchpad, not a Landing

F or many, the word "retirement" conjures images of quiet days spent relaxing. But for some of us, it is less a gentle landing and more a brand-new launchpad for exploration and contribution. Retirement isn't about sitting around being bored; It is about repurposing experience and passion to make a difference in new and exciting ways. At least, that's how It is been for me since I wrapped up my executive career at NASA.

Just months after leaving NASA in 2016, I found myself joining the Johns Hopkins Applied Physics Lab, lending my expertise to independent investigations and proposal reviews. But beyond the professional pursuits, a strong desire to give back truly took center stage.

I've always felt that when you've held a high-level government position, it is important to share what you've learned. My passion for STEM+AD (Science, Technology, Engineering, Math plus Arts and Design) education led me straight into classrooms. I loved inspiring middle and high school students with tales of my journey to NASA, hoping to ignite a spark in their young minds.

One particular encounter really stuck with me. A young girl in one of the classes that I was a guest speaker in doubted her own potential, saying she could not achieve much, and it resonated deeply, reminding me of my own struggles back in high school. That moment fueled my desire even more to empower young minds and help them find their confidence.

Then came the COVID pandemic, and I took on an unexpected role: instructing a semester of Aerospace Engineering to junior and senior high school students. Most classes were online, of course, but we managed to discuss key topics in person when we could. I brought in some amazing guest speakers via online: a Project Manager and Program Scientist who shared essential skills and real-time examples from their work; an astronaut who recounted his incredible experiences in space; even a Nobel Laureate who discussed his initial failures and eventual success through sheer tenacity and perseverance, and a cybersecurity expert who spoke about her educational challenges and the significance of cyber security today. Determined to provide a memorable experience despite the isolation of online learning, I was incredibly gratified by the heartwarming feedback I received from the students at the end of the semester.

I challenged them with a paper exercise design project: build a Lunar hopper. The goal was to focus on objectives, design requirements, including environmental constraints, and design the hopper. I left plenty of room for creativity but secretly included one unattainable requirement that over-constrained the design, without the students knowing the implications. They only discovered this challenge during our final in-person lesson! When they realized their design couldn't meet the requirements, we discussed real-life issues of unmet performance expectations and the need to reevaluate constraints. This "fail without failing" exercise was actually inspired by a previous discussion I had with astronaut Kathy Sullivan, who emphasized the importance of teaching students about persistence despite setbacks. At the end of the semester, students told me it was the most memorable lesson of all.

That whole experience really deepened my respect for educators. I witnessed firsthand that teaching demands significantly more time than is typically allotted for a single class. It is truly a labor of love!

Beyond the classroom, I joined the local school district's Citizen Financial Advisory Group, where I contributed my financial management expertise. And I answered the call to serve my community as a Town Commissioner, applying my knowledge to benefit local residents.

In 2019, NASA sought my leadership again, appointing me to lead the independent review team for the Volatiles Investigating Polar Exploration Rover (VIPER) mission to the Moon. VIPER would have analyzed ice and other volatiles on the lunar surface and subsurface, providing crucial data for future lunar missions under the Artemis program. The VIPER project was ultimately canceled due to funding issues, both with the rover and the industry-provided lander. Simultaneously, I co-led an independent review of NASA's Earth Observatory System plans, ensuring the effective coordination of Earth science satellites.

It is clear that retirement hasn't meant slowing down. The shift to remote work during the pandemic allowed me to pursue these various initiatives while still prioritizing family time. Speaking of which, a memorable trip to Las Vegas with my wife, Lauretta, led to an incredible Blue Man Group performance filled with surprises. A friend in Vegas arranged the tickets for us, but there was a special surprise in store.

We were told to arrive early for the show and were seated halfway back in the auditorium with an aisleway directly behind us. We were a little confused, wondering why our friend's "connections" didn't get us better seats. But then, just before the show started, an usher came and asked us to follow him. As we walked across the aisle, a spotlight shone on us, and the music played, "You're late, you're late, you're holding up the show, you're late." It was embarrassing, to say the least! We were then led to center seats, eight rows from the front. Later, we found out those closer than eight rows would get splattered with water during the show. After the performance, an usher kindly gifted us a pair of drumsticks used during the show. One of the Blue Man Group members even autographed the sticks for us, leaving his signature blue lip prints on them! It was such an exciting adventure!

After the Blue Man Group performance

Since then, Lauretta and I have embarked on numerous adventures, exploring new destinations and revisiting cherished family moments. From cruises through the Alaskan inner passage, the Hawaiian Islands, and the Panama Canal.

Now, I'm consciously dedicating more time to leisure activities, school engagements with students, focused family time, and scaling back on professional commitments to truly enjoy the fruits of my labor. However, my passion for space exploration and NASA's contributions remains unwavering. I hosted Delaware Congressman Brian Pettyjohn at the Wallops Flight Facility for an

Antares launch event, showcasing the vital work conducted there.

In the ongoing balancing act of life, my deepest desire is to carve out more meaningful time with family. It is a commitment I strive for amidst the responsibilities of prioritizing mentoring roles and actively engaging in school activities. Yet, it is often during specific moments that this desire truly crystallizes, like watching our grandson on the ice. His passion for hockey is infectious, and seeing him with his sleek, modern composite stick, executing those powerful slap shots, undeniably transports me back to my own teenage years on frozen ponds. Those were simpler days, fueled by raw enthusiasm and the biting cold, where the "arena" was a snow-banked patch of ice. Our tools of choice were vastly different: sturdy, weighty wooden hockey sticks. There was no talk of "flex points" or "kick points" then; you felt every bit of the puck on the blade, and power came from pure upper-body strength and technique. Compared to our grandson's lightweight, high-performance stick, engineered for incredible whip and shot velocity, it feels like an entirely different sport. The technological leap, especially in the stick's flex and material, is truly astounding—a testament to how the game has evolved, yet the core joy of the game, the camaraderie, and the simple thrill of a well-placed shot remain timeless. These shared moments, whether on the pond or in a heated (still cold) arena, are the tapestry of our family life, reminding me why these connections are the most important investments of all.

These stories are a testament to the fact that retirement is not an ending, but a new chapter. It is an opportunity to contribute, explore, and continue learning, all while cherishing the people and passions that bring joy and fulfillment. It is a new launchpad, ready to propel you toward new and exciting horizons.

Chapter Seventeen

Science and Faith – The Enduring Spirit

For many, the name NASA conjures images of soaring rockets, distant galaxies, and groundbreaking discoveries. And rightfully so. For me, it represents something more profound: a tireless pursuit of knowledge, a dedication to innovation, and a profound human yearning to understand our universe. I've been incredibly honored to be a part of this remarkable organization, contributing in my own way to its legacy of scientific and technological advancement.

It is about being surrounded by brilliant minds, all driven by a common goal: to expand the limits of human understanding. Whether a mission focuses on unraveling the mysteries of the universe through sophisticated telescopes or meticulously studying the impacts of climate change on our planet, the underlying motivation remains the same—to seek answers to fundamental questions. This yearning for answers resonates deeply within me. The inherent human desire to explore the unknown and probe the unexplainable is what fuels innovation and drives us forward. It is this spirit that permeates every corner of NASA, from the engineers designing cutting-edge spacecraft to the scientists analyzing data collected from light-years away.

Working alongside these individuals has been an invaluable learning experience. Witnessing firsthand the meticulous planning, countless hours of dedicated labor, and the unwavering commitment to excellence that goes into each mission is truly inspiring. The collaborative spirit, the willingness to share knowledge, and the constant encouragement to push boundaries create an environment where innovation truly thrives. It is an ecosystem of shared purpose, where every challenge is met with collective ingenuity and persistent optimism.

Whether my contribution has been small or large, knowing that I've played a part in this grand endeavor fills me with immense pride. Being a part of NASA is not just a job; It is an opportunity to contribute to something bigger than oneself, something that will inspire generations to come. It is a chance to be part of history, to help shape our understanding of the universe, and contribute to solving the challenges facing our planet. Each day brings a renewed sense of purpose, knowing that our collective efforts contribute to a future where humanity's grasp of the cosmos continues to grow.

The future holds endless possibilities for scientific discovery and technological advancement. And I am grateful to have had the opportunity to be a part of the journey, pushing the boundaries of knowledge with an organization that embodies the very essence of human curiosity and ingenuity. The yearning to seek answers will continue to drive us forward, and I am excited to see what the future holds for NASA and for the world. Our journey of discovery is far from over, and the stars continue to beckon us onward.

As an engineer, trained to objectively analyze information, I've often pondered how the relentless pursuit of scientific knowledge can sometimes appear to conflict with faith. How does one reconcile the breathtaking wonders of the universe with deeply held religious beliefs? I've witnessed firsthand the vibrant debates and disagreements among scientists interpreting data, and I've come to see that such discussions are not only necessary but beautiful. Science, by its very nature, is an evolving system of knowledge, ever shifting and refining our understanding. The journey from a geocentric model during the second century, where they believed the sun revolved around the Earth, to a heliocentric model in the sixteenth century, where the Earth and planets revolved around the sun, or the continuous revelations from observatories like the Hubble and James Webb Space Telescopes, beautifully highlight this dynamic process. This beautiful incompleteness demands humility in our conclusions.

For me, scientific exploration and faith are not mutually exclusive; in fact, they profoundly strengthen one another. When I witness the awe-inspiring images beamed back from distant observatories, I am filled with an overwhelming sense of wonder. Just as we use science to understand the human body's intricate systems and create antidotes for illness, we use science to better understand our universe. The more we learn, the more profound the creation appears.

I find immense comfort and resonance in the words of the Psalmist in Psalm 96:11-12 (NIV): "Let the heavens be glad, and the earth rejoice! Let the fields be jubilant, and everything in them; let all the trees of the forest sing for joy." When I see nature in action, when I marvel at the majestic harmony of the universe revealed through scientific eyes, I see exactly what the Psalmist wrote about. My journey with NASA, exploring the vastness above and the intricacies within, does not conflict with my faith but rather deepens my belief in a higher being. I eagerly anticipate images from missions not yet planned, knowing each new discovery offers another glimpse into the boundless wonder of our existence.

I believe we are all endowed with distinct gifts, much like the different parts of a body working in harmony. For me, one of those gifts is compassion. This innate quality compels me to treat every individual with profound respect, regardless of their beliefs or background. It is a simple philosophy: I strive to treat others as I would wish to be treated. To me, living a life that genuinely inspires others is the most authentic expression of my faith in action. It is about building bridges, not walls, and finding common ground in our shared humanity.

The journey of faith, I've come to understand, is deeply personal. The choice to embrace a particular belief system belongs solely to the individual, and I truly believe that only God can truly touch a heart and guide it toward understanding. My role, as I see it, is not to persuade or convert, but to embody the principles I hold dear.

I'm often reminded of a profound discussion I had with my late brother, a conversation that deeply shaped my perspective on faith and behavior. He concluded our exchange with a statement that has resonated with me ever since:

"Consider this," he began. "What if I am wrong about my faith, and there is no God, no heaven, and no hell? Even in that scenario, living a life as if there is a God—meaning maintaining integrity, treating others with respect, helping those in need, and striving for a life of no regret—is inherently valuable. It is a win-win, regardless of your belief."

He paused, then continued, "But what if I am right? What if there is a God, and there are a heaven and hell? And if I have not accepted Christ, I will live in eternity in hell. Is that a gamble you truly want to take?"

His words were not meant to instill fear, but to provoke thought, to highlight the profound implications of our choices and how we live our lives. This conversation reinforced my conviction that how we conduct ourselves, day in and day out, is paramount. It is about more than just belief; It is about embodiment. Whether we subscribe to a particular faith or not, cultivating a life of integrity, compassion, and purpose is a journey worth undertaking—a testament to the best of what we can be as human beings. It is a message I carry with me, inspiring me to live in a way that truly matters, both to myself and to those around me.

Chapter Eighteen

Reflections on My Career

Y ou know, it is funny how all those little things from our past, good and bad, sneakily shape who we become. For me, a small-town farm kid from Pennsylvania, those early influences, mixed with a good bit of grit and a lifelong commitment to service, somehow led me to an absolutely wild and extraordinary career at NASA.

Reflection on my journey - from childhood to NASA

I completely understand what many young people I have met feel like. I, too, used to feel like I was not good enough to achieve anything, and honestly, pretty alone sometimes. But once I started embracing my adventures and pushed past those early doubts and negative opinions, things really began to change. Now, I'm deeply committed to passing on that hard-earned wisdom to students, hoping to encourage them to chase their dreams without letting anyone else put limits on them.

My journey really kicked off on that farm. That is where I learned some very invaluable lessons about honesty, integrity, and taking responsibility. Those principles have become the foundation of who I am, and honestly, they still guide nearly everything I do today. Even simple childhood activities, like building model rockets, taught me a deep respect for rules and an understanding of consequences. Funny how that works, right?

Throughout my career, I've tried to live by a straightforward philosophy: always try to find the good in people, offer encouragement whenever it is needed, and approach decisions with a good dose of thoughtful deliberation. Patience, faith, and good leadership have been my unwavering guiding lights. While treating others with respect is always paramount, I've also come to realize how important it is to face challenges head-on—that's how you really grow.

The most rewarding part of my work, without a doubt, has been seeing the success of those I've mentored. Upholding integrity and tackling problems directly have always been my top priorities. And consistency in action? That's crucial. I always understood that others were watching, so I tried my best to walk the talk.

I used to have a routine I followed: every Friday morning, I would set aside some time just to think about the week's events. I would reflect on people who stood out to me that week and then craft a short appreciation email to them. If they were in the same building as me, I would often just pop over to their office and tell them in person. It is a trait I actually picked up from a great mentor early in my career. It really makes a difference.

I vividly recall the coworker who encouraged me to attend community college—that was a truly pivotal moment that eventually propelled me toward my whole NASA career. I'm sure Lloyd has no idea how much of an impact he has had on me. It really makes you think about how we all touch someone's life without even knowing the ripple effect we have.

Recognizing the lack of support I experienced during my own high school years, I'm now truly committed to mentoring students. I want to guide them through their challenges and hopefully prevent them from going through some of the struggles I did.

My ultimate achievement—rising to lead the largest civilian space science organization in the world as a Senior Executive Service member—really feels like a testament to my dedication and resilience. When I look back, my life experiences truly fit into this timeless principle: "Follow the Golden Rule: Do Unto Others as You Would Have Them Do to You." It just works.

A health scare really made me pause and reflect on my future, and ultimately, it led to the decision to step down. I realized that my legacy would extend way beyond my professional achievements. I felt a strong desire to share my experiences, to inspire others to persevere, and, honestly, to dedicate more time to my family and mentorship.

I often think about those stunning Hubble images of the Carina Nebula, which use both optical and infrared technology. The optical view is similar to what you would see looking through a glass or reflective-based telescope without modifications. The gases, as seen in these magnificent gas pillars, mask the countless stars and galaxies deeper in the cosmos. The infrared instruments are designed to penetrate through these gases, revealing the beauty beyond the pillars of gas: the countless stars and galaxies. It brings home the importance of looking beyond surface appearances.

There's almost always more information than initially apparent, so it is always best to hold off on drawing premature conclusions.

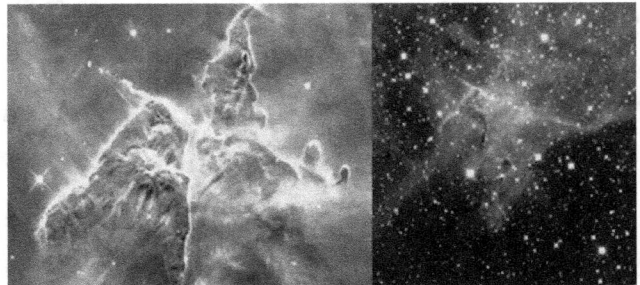

Carina Nebula with visible and IR Spectrum views: Credit NASA, ESA, and M. Livio and the Hubble 20th Anniversary Team (STScI)

And I must say that none of my successes would have been possible without the unwavering support of my amazing wife, Lauretta. She's been my rock.

To stay grounded amidst the complexities of such a demanding career, I put together a list of personal mantras that I kept at my desk. Things to remind me to be present, grateful, and not to sweat the small stuff. I also kept a journal of key events, meaningful phrases, and advice from the mentors who shaped me—it was a constant source of inspiration and direction.

Ultimately, I always returned to my core value: "Maintain uncompromising integrity." This principle, along with all the countless lessons learned from a life lived among the stars, will continue to guide me as I embark on my next chapter, dedicated to empowering the next generation of dreamers and achievers. It is going to be a fun ride.

Key Leadership Attributes to Follow (the list I had on my computer monitor)

Need new goals for each new day

SENSE OF PURPOSE

SPIRIT OF ADVENTURE

CAPacity for growth

Develop characteristics

UNCOMPROMISING INTEGRITY

WORKING PRIORITIES (NOT ONLY DEVELOPING THEM)

BLINDING FLASH OF THE OBVIOUS

Start with number one on the priority list and continue on priorities

Expression: If you have a frog to swallow, don't look at it too long. If you have two, swallow the biggest one first.

Courageous—either you or your fear takes control.

Expression: Don't wait until all the stoplights are green before going across the city.

Start gaining altitude before making a turn

A crisis must never be experienced a second time

Goal orientation: ability to make tough decisions

Don't get frustrated—sometimes change is needed

Expression: You can't teach a pig to sing; it wastes your time, and it makes the pig mad trying

Inspired enthusiasm

Levelheaded—great speed in grasping the facts

Ability to select and develop good people—hire ones better than you

There is no situation so bad that a single action cannot make it worse.

4 Key Themes

1. Dream

2. Study everything you get your hands on

3. Plan your time, time your plan (It is ok to have a tiger by the tail if you know what to do next)

4. May you never be less than your dreams

5 to 1 Rule

PROVIDE A MINIMUM OF 5 COMPLIMENTS TO 1 CRITICISM.

3 Out of 5 Rule

ON AVERAGE, 3 OUT OF 5 DAYS A WEEK, FEEL GOOD ABOUT WHAT WAS ACCOMPLISHED, BELIEVE YOU CONTRIBUTED TO THE MISSION, AND WANT TO RETURN THE NEXT WEEK. ANYTHING SHORT OF "3 OUT OF 5" INDICATES A WRONG JOB ASSIGNMENT OR BEING IN THE WRONG POSITION.

Keep your head above the clouds and your feet firmly planted on the ground—then move step by step.

Life Lessons Etched in Gratitude

Life, in its beautiful and often chaotic way, leaves indelible marks on our souls. Each moment, each person we encounter, each lesson learned, contributes to the tapestry of who we become. These experiences are deeply personal treasures, carried close to the heart and forever cherished. Reflecting on these formative moments, I'm particularly struck by the profound impact an educator can have on a young life. My own trajectory was significantly altered by a 5th-grade teacher who recognized and nurtured my potential long before I saw it myself. Their belief and unwavering support served as a catalyst, setting the stage for a future I hadn't dared to imagine. It underscores the vital role educators play in shaping the leaders of tomorrow, and for those who saw something in me, I am eternally grateful.

And so, I share with you the life lessons woven from these experiences, hoping they will inspire you to reach for your own stars, or to encourage a young person to do the same:

Lesson #1: Find your own safe space

In the ceaseless demands and interconnectedness of modern life, it is remarkably easy to feel stretched thin, overwhelmed by responsibilities, constant digital input, and the sheer volume of external noise. That's precisely why, sometimes, the most crucial antidote isn't more stimulation, but simply the profound gift of your own space. This isn't about isolation, but rather about creating a personal sanctuary where the mind can finally declutter, the nervous system can settle, and the spirit can breathe freely. Within this quiet refuge, you can process thoughts, calm anxieties, and reconnect with your inner self, allowing you to truly "recharge your batteries" after depletion. This period of solitude isn't luxury; It is a foundational act of self-care that empowers you to return to the world with renewed energy, sharper focus, and a greater capacity to navigate its challenges.

Lesson #2: Let doubts and criticisms become your driving force that propels you forward

The beauty of the human spirit lies in its resilience and ability to defy expectations. Instead of internalizing these negative pronouncements, we can choose to use them as fuel for growth. Let the doubts and criticisms become the driving force that propels you forward.

Lesson #3: Embrace the unexpected

Sometimes, the detours we take, fueled by a refusal to accept limitations, lead us to discover passions and talents we never knew we possessed. Don't let anyone write your story for you. Pick up the pen and write it yourself. You might just surprise yourself with the incredible chapters you create.

Lesson #4: Life is about finding the right balance

Because, just like adjusting the water injection rate on a Ford Escort, life is all about finding the right balance. Life, at its very core, is a perpetual quest for balance. It is the delicate interplay between arduous work and rejuvenating rest, the mindful allocation of time between cherished relationships and necessary solitude, and the constant negotiation between ambition and contentment. Without this harmony, we risk burnout, emotional exhaustion, or a pervasive sense of disconnect. The right balance isn't a fixed state, but a dynamic, ever-adjusting journey, demanding constant awareness and adaptation as life's demands shift. Mastering this art of equilibrium allows us to navigate challenges with resilience, savor joys more deeply, and cultivate a sustainable sense of inner peace and fulfillment.

Lesson #5: Focus on the good in people

It is human nature to notice flaws, to point out imperfections. But one of the most valuable lessons I've learned is the transformative power of focusing on the good in others. Shifting your perspective from judgmental to empathetic, from pessimistic to optimistic, can have a profound impact on your interactions and overall outlook. When you actively seek out the positive qualities in those around you, you'll be amazed by the uplifting effect it has. Remember, everyone is navigating their own unique struggles and strengths. By fostering understanding and compassion, we contribute to a more harmonious and supportive world.

Lesson #6: Offer support and mentorship to those who need an extra boost

We all encounter moments when we need a helping hand, a word of encouragement, or simply someone to believe in us. Imagine a friend or colleague brimming with untapped potential yet somehow held back from achieving their goals. By offering your support, guidance, and a listening ear, you can help them navigate challenges and unlock their hidden capabilities. Don't hesitate to lend a hand; together, we can empower others to achieve extraordinary things.

Lesson #7: Actions have consequences

This is a simple truth, yet one easily forgotten in the heat of the moment. Every action, big or small, carries consequences, whether positive or negative. Before making a major decision, take a moment to consider the potential ramifications. A little foresight can save you from future regret.

Lesson #8: Experiences shape the attitude

Life often throws us curveballs, forcing us to confront unforeseen challenges and profound loss. These experiences, although painful, can profoundly reshape our perspective. I learned to set aside my own grief to support others, cultivating resilience, recognizing the quiet strength found in community, and appreciating the enduring human capacity to carry on, even when the weight feels unbearable. These tough times often become blessings in disguise, pushing us to grow and evolve in ways we never imagined possible, ultimately leading to a better appreciation for life's preciousness.

Lesson #9: Patience is a virtue

Impulsivity can be tempting, especially when faced with a difficult decision. However, learning to cultivate patience is crucial. Take a step back, gather all the facts, and consider the implications carefully before making a move. Rushing into something without fully understanding the situation can lead to negative outcomes. Waiting for more information or seeking advice from trusted sources can often lead to far better results. Your future self will thank you for exercising patience and making an informed decision.

Lesson #10: Sometimes the most unexpected moments lead to the most rewarding connections

The next time you find yourself in an embarrassing situation, remember the red wine in Paris. Picture that moment not as a stumble, but as an elegant, albeit messy, dance step in the grand ballroom of life. Embrace the vulnerability that often accompanies such mishaps; It is a raw, human quality that disarms and invites authenticity. Find the humor, for laughter is the most graceful way to navigate awkwardness, transforming blushes into smiles and shared understanding. Recognize that this very moment, however cringeworthy it may feel, might just be the catalyst for something truly extraordinary—a deep connection formed over shared laughter, an insightful realization about your own resilience, or even the opening of a door you never knew existed. After all, sometimes the most unexpected, unscripted, and even awkward moments lead to the most rewarding connections and the most memorable stories.

Lesson #11: Walk through open doors

Sometimes, life presents us with opportunities shrouded in uncertainty. It can be daunting to take a leap of faith without all the details perfectly mapped out. Yet, it is often in these moments of uncertainty that we discover incredible adventures we would never have imagined. By embracing new experiences and taking a chance, even when things aren't crystal clear, we open ourselves up to growth and unexpected blessings. Who knows what incredible journeys await on the other side?

Lesson #12: Maintain a standard of high integrity

Perhaps the most important lesson I learned growing up on the farm was the importance of integrity. Take responsibility for your actions, honor your commitments, and conduct yourself with honesty and ethical behavior. It is about holding ourselves accountable for our mistakes, standing by our words, and following through on our promises, regardless of the circumstances. Being a person of integrity means doing the right thing even when no one is watching, staying true to your values and beliefs. It is about being honest, ethical, and trustworthy in everything you do. Whether it is showing up on time for a meeting or admitting when you've made a mistake, taking responsibility, honoring commitments, and maintaining high integrity are key ingredients to building strong relationships and earning the trust of others.

Lesson #13: Be careful about making hasty assumptions

In almost any situation, there is invariably more information to uncover than what initially meets the eye, making it paramount to resist the urge to draw premature conclusions. What appears on the surface is often just a symptom, with deeper, interconnected factors, subtle nuances, or alternative perspectives waiting to be discovered. Leaping to hasty judgments based on limited data inevitably leads to misunderstandings, propagates misinformation, and results in flawed decisions that can have lasting, negative repercussions. Conversely, cultivating patience and committing to further inquiry allows for a more comprehensive understanding, revealing the broader context, identifying critical details, and ultimately fostering clarity and accuracy. This disciplined approach of reserving judgment until a more complete picture emerges is not just cautious; it is a mark of wisdom, enabling genuinely informed choices and fostering a more nuanced and just understanding of any given situation.

These lessons, etched in the fabric of my life, serve as a constant reminder of the power of human connection, the importance of integrity, and the potential for growth that lies within each of us. May they inspire you to live a life filled with purpose, compassion, and an unwavering belief in the goodness of humanity.

Chapter Nineteen

Final Thoughts

The Whisper of the Cosmos: Navigating the Fleeting Moment with an Eternal Heart

There is a profound, almost paralyzing irony to the human experience: we are molded by the infinite yet constrained by the immediate. We wrestle with bills, deadlines, and the fleeting opinions of others, often forgetting the scale against which our struggles are measured. If we could look beyond our time here on earth, our lifespan would appear as nothing more than a short blip in time.

This realization is not meant to incite panic or despair, but rather to lend urgency to every rising sun. Life is not a dress rehearsal; it is the single, decisive performance that sets the stage for everything that follows.

The Hourglass and the Eternal Sea

We live in a culture that simultaneously cherishes immediacy and permanence, striving ironically for eternal fame while fixating on instant gratification. Yet, the great teachers and prophets throughout history have consistently reminded us: nothing of this world is truly permanent save for the state of our soul and the residue of selfless love we leave behind.

Consider the span of a thousand years, a period so vast it renders monuments into dust and empires into footnotes. Now try to imagine eternity—not just a long time, but *timelessness* itself. When you think about eternity, our eighty or ninety years on Earth are reduced to the flicker of a match in a darkened cathedral.

This thought demands a shift in perspective. If the journey is short, we pack only what is essential. If the journey determines our final resting place, we must choose our direction with absolute clarity.

We understand the human tendency to drift. The constant distractions—the demands of career, the allure of comfort, the song of self-focus—are powerful, pulling us effortlessly away from intentional living. It is easy to spend decades building a beautiful house on temporary ground, only to realize too late that the foundation was built on sand.

The Weight of Today's Small Choices

The greatest challenge in grasping eternal consequences is that eternity rarely manifests in grand, visible moments. It is not always about the dramatic conversion or the public sacrifice. More often, *the eternal consequences are woven into the quiet, repetitive fabric of our daily choices.*

Every choice we make—how we respond to irritation, whether we choose patience over anger, integrity over convenience, generosity over hoarding—is not merely a transient action. It is an act of spiritual self-sculpting. We are, moment by moment, forging the very nature of the self we will carry into eternity.

If we consistently choose callousness, we harden our eternal heart. If we consistently choose forgiveness, we refine our soul's capacity for grace. The everlasting consequence is thus not solely a reward *given* to us, but the fully realized, perfected (or regrettably stunted) version of the self we have allowed God to shape us during our time on Earth.

This burden of choice can feel heavy, even overwhelming. We are inherently fallible; we stumble, we regress, and we often choose poorly out of weakness or fear. The world tells us that if we fail, we are defined by that failure. But the truth of the divine plan is far more compassionate: what defines us is not whether we fall, but in which direction our hearts are ultimately leaning, and whom we allow to guide our recovery.

The Imperative of the Guiding Light

In our journey here on Earth, we need an anchor, a compass, and a light to cut through the fog of worldly priorities. We cannot afford to navigate this pivotal journey relying solely on our own fallible intellect or rapidly shifting emotions.

We need a guide who has already traversed the path, who understands the treacherous turns of the human heart, and who loves us despite our predictable failings.

May Christ Be Your Guiding Light

This is not an empty spiritual cliché; it is the singular, practical imperative for making wise choices with eternal significance.

To choose Christ as the guiding light means choosing *the Way* He demonstrated: a path defined not by power or temporal success, but by self-giving love. It means choosing *the Truth* He proclaimed, which offers clarity in a world saturated with half-truths and manufactured realities. It means choosing *the Life* He offers, a life of grace that forgives our stumbles and empowers us to rise and choose rightly next time.

When faced with a complex choice that seems to waver between worldly success and spiritual integrity, Christ offers the ultimate simplification: *Choose love*. Choose the path that reflects the sacrificial posture of the cross.

His guidance does not remove the difficulty of the choice, but it alters the outcome. When we align our will with His, our short journey becomes infused with eternal significance. Our efforts shift from self-preservation to soul-salvation, not just our own, but that of those around us.

Choose Wisely, Live Intentionally

We are travelers in a temporary land, and the time for hesitation is past. Let us pause today, amidst the noise and the haste, and look across the vast spectrum of eternity. Let that vision clarify our immediate tasks.

Let us recognize the unique honor and terrifying responsibility of this brief lifespan. We are living in the moment that determines the infinite.

Therefore:

PRIORITIZE THE ETERNAL: INVEST YOUR ENERGY, TIME, AND HEART IN VIRTUES—FAITH, HOPE, AND ABOVE ALL, LOVE—THAT WILL SURVIVE THE GRAVE.

EMBRACE GRACE: DO NOT LET PAST FAILURES PARALYZE YOU. TURN TOWARD THE LIGHT; ACCEPT THE OFFERED FORGIVENESS AND START ANEW, TODAY.

FOLLOW THE LIGHT: IN EVERY DECISION, GREAT OR SMALL, ASK YOURSELF: DOES THIS CHOICE REFLECT THE HEART OF CHRIST? DOES IT MOVE ME CLOSER TO THE ETERNAL HOME HE PROMISED?

Your life is a whisper in the cosmos, but it is a decisive whisper. Choose wisely, my fellow traveler, and allow the unwavering, compassionate love of Christ to guide you, transforming your short journey on earth into a prelude of eternal glory.

Be the author of your own story!

Chapter Twenty

References

ESA mission

More information related to the European Space Agency Rosetta (rendezvous with a comet) mission can be found at https://rosetta.esa.int

Hubble Space Telescope

Hubble Space Telescope launch in 1990 has changed the fundamental understanding of the universe. The telescope experienced an anomaly where the imagery focus was not optimum, requiring correction. NASA conducted a special mission to the HST to correct the focus issue. A total of 5 repair and upgrade missions were conducted on the HST, resulting in Imaginary from the HST that is nothing less than breathtaking, enabling scientific discoveries that could previously only be theorized.

More information related to the Hubble Space Telescope can be found at https://science.nasa.gov/mission/hubble/

James Webb Space Telescope

The James Webb Space Telescope (JWST), launched on December 25, 2021, is the scientific successor to the Hubble Space Telescope and complements and extends the discoveries of the Hubble Space Telescope.

More information related to the JWST can be found at https://science.nasa.gov/mission/webb/

Juno

Juno was launched on August 5, 2011, traveling nearly five years and 1.7 billion miles, evading showers of the most punishing radiation outside the Sun before reaching Jupiter and entering Jupiter's orbit on July 5, 2016.

More information related to the Juno mission can be found at https://science.nasa.gov/mission/juno/#more-about-the-mission

NASA Educational services

https://www.nasa.gov/learning-resources/

NASA SMD Science Fleet

More information related to the SMD fleet of mission can be found at

https://science.nasa.gov/science-missions/

OSIRIS-REx

The mission was launched on September 8, 2016, traveling to a near-Earth asteroid named Bennu. One of the mission objectives was to collect a sample of rocks and dust from Bennu's surface and return the samples to Earth. The samples were collected on October 20, 2020, and delivered to Earth on September 24, 2023.

More information related to the OSIRIS-REx mission can be found at https://science.nasa.gov/mission/osiris-rex/

Pluto New Horizons

New Horizons launched January 19, 2006, was the first spacecraft to explore Pluto up close, flying by the dwarf planet and its moons in 2015.

https://science.nasa.gov/mission/new-horizons/

Senior Executive Service (SES)

Members of the SES serve in the key positions just below the top Presidential appointees. SES members are the major link between these appointees and the rest of the Federal workforce. They operate and oversee nearly every government activity in approximately 75 Federal agencies

More information related to SES can be found at https://www.opm.gov/policy-data-oversight/senior-executive-service/

Volatiles Investigating Polar Exploration Rover (VIPER)

The VIPER is designed to explore the relatively nearby but extreme environment of the Moon in search of ice and other potential resources. This mobile robot was slated to land at the South Pole of the Moon on a 100-day mission. The critical information it provides will teach us about the origin and distribution of water on the Moon and help determine how we can harvest the Moon's resources for future human space exploration. As of this writing, the VIPER mission was canceled.

More information related to the VIPER mission can be found at https://science.nasa.gov/mission/viper/